Modeling, Sensing and Control of Gas Metal Arc Welding

Full catalogue information on all books, journals and electronic products can be found on the Elsevier Science homepage at: http://www.elsevier.com

ELSEVIER PUBLICATIONS OF RELATED INTEREST

JOURNALS:

Advanced Engineering Informatics
Annual Reviews in Control
Automatica
Computers and Electrical Engineering
Computers in Industry
Control Engineering Practice
Engineering Applications of AI
Image and Vision Computing
International Journal of Heat and Mass Transfer
International Journal of Machine Tools and Manufacture
International Journal of Pressure Vessels and Piping
International Journal of Machine Tool and Manufacture
Journal of Manufacturing Systems
Journal of Materials Processing Technology
Journal of the Franklin Institute
Measurement
Mechatronics
Robotics and Autonomous Systems
Robotics and Computer-Integrated Manufacturing
Sensors and Actuators
Signal Processing

Modeling, Sensing and Control of Gas Metal Arc Welding

Desineni Subbaram Naidu
Idaho State University
Idaho, USA

Selahattin Ozcelik
Texas A&M University
Texas, USA

Kevin L. Moore
Utah State University
Utah, USA

2003
ELSEVIER
Amsterdam – Boston – London – New York – Oxford – Paris – San Diego
San Francisco – Singapore – Sydney – Tokyo

ELSEVIER SCIENCE Ltd
The Boulevard, Langford Lane
Kidlington, Oxford OX5 1GB, UK

© 2003 Elsevier Science Ltd. All rights reserved.

This work is protected under copyright by Elsevier Science, and the following terms and conditions apply to its use:

Photocopying
Single photocopies of single chapters may be made for personal use as allowed by national copyright laws. Permission of the Publisher and payment of a fee is required for all other photocopying, including multiple or systematic copying, copying for advertising or promotional purposes, resale, and all forms of document delivery. Special rates are available for educational institutions that wish to make photocopies for non-profit educational classroom use.

Permissions may be sought directly from Elsevier Science & Technology Rights Department in Oxford, UK: phone: (+44) 1865 843830, fax: (+44) 1865 853333, e-mail: permissions@elsevier.com. You may also complete your request on-line via the Elsevier Science homepage (http://www.elsevier.com), by selecting 'Customer Support' and then 'Obtaining Permissions'.

In the USA, users may clear permissions and make payments through the Copyright Clearance Center, Inc., 222 Rosewood Drive, Danvers, MA 01923, USA; phone: (+1) (978) 7508400, fax: (+1) (978) 7504744, and in the UK through the Copyright Licensing Agency Rapid Clearance Service (CLARCS), 90 Tottenham Court Road, London W1P 0LP, UK; phone: (+44) 207 631 5555; fax: (+44) 207 631 5500. Other countries may have a local reprographic rights agency for payments.

Derivative Works
Tables of contents may be reproduced for internal circulation, but permission of Elsevier Science is required for external resale or distribution of such material.
Permission of the Publisher is required for all other derivative works, including compilations and translations.

Electronic Storage or Usage
Permission of the Publisher is required to store or use electronically any material contained in this work, including any chapter or part of a chapter.

Except as outlined above, no part of this work may be reproduced, stored in a retrieval system or transmitted in any form or by any means, electronic, mechanical, photocopying, recording or otherwise, without prior written permission of the Publisher.
Address permissions requests to: Elsevier Science & Technology Rights Department, at the phone, fax and e-mail addresses noted above.

Notice
No responsibility is assumed by the Publisher for any injury and/or damage to persons or property as a matter of products liability, negligence or otherwise, or from any use or operation of any methods, products, instructions or ideas contained in the material herein. Because of rapid advances in the medical sciences, in particular, independent verification of diagnoses and drug dosages should be made.

First edition 2003

Library of Congress Cataloging in Publication Data
A catalog record from the Library of Congress has been applied for.

British Library Cataloguing in Publication Data
A catalogue record from the British Library has been applied for.

ISBN: 0 08 044066 5

∞ The paper used in this publication meets the requirements of ANSI/NISO Z39.48-1992 (Permanence of Paper). Printed in The Netherlands.

Dedicated to

Herschell B. Smartt

for introducing this research topic to us.

- - - *Authors*

PREFACE

Arc welding is one of the key processes in industrial manufacturing. It is believed that in the entire metal fabrication industry, arc welding is the third largest job category behind assembly and machining. In particular, two types of processes, gas metal arc welding (GMAW) and gas tungsten arc welding (GTAW), are extensively used in factory automation. The practice of industrial welding is heavily dependent on the knowledge and vast experience of the welder, and, as such, at present it is more an art than science. However, in order to make the welding process more *automated* and less human (welder) dependent, in the last two decades significant efforts have been made to introduce the ideas of feedback in order to *control* the welding process to achieve a *good* weld. Modeling a welding process is one of the most important ingredients for controlling the process. Various modeling techniques and automatic control strategies have been suggested to improve the welding process.

In this research monograph we provide a survey of modeling, sensing, and automatic control of the GMAW process. As such, this volume is intended to be one of *survey-oriented* rather than a detailed account of the subject matter. Further, the references under each chapter are listed chronologically so that one can see the development of the field over the years. To date, it appears that there have been at least three related literature surveys: one in 1988 on *expert robotic* welding systems (Sicard and Levine 1988), another in 1989 on sensory feedback control for *robotic* arc welding (Cook 1983a), and a third tutorial type of survey in 1989(Cook et al. 1989a). The first two focused on only robotic welding, while the third considered modeling, sensing and automatic control of the GMAW process. The main purpose of the present monograph is to provide researchers with an updated status of the state-of-the-art in the areas of modeling, sensing and automatic control of the GMAW welding process. In addition to survey, this monograph also collects a number of original research results on this topic by the authors and their colleagues.

This volume is composed of 6 chapters. After introducing some preliminary classification of various welding processes in Chapter 1, the modeling aspects GMAW process is briefly discussed in Chapter 2. The topics reviewed are physics of welding, metal transfer characteristics, weld pool geometry, process voltages and variables, power supplies, etc. The topic of sensing is reviewed in Chapter 3 touching upon various sensors for arc length, weld penetration control, weld pool geometry, and optical and intelligent sensors. The main theme of control is dealt with in Chapter 4 starting from the classical control techniques of PI, PID, multivariable control, adaptive control, and intelligent control. A special feature of this book is in Chapter 5 where a case study is presented by the authors and their students at Idaho State University (ISU), Pocatello, Idaho in collaboration with the researchers at the Idaho National Engineering and Environmental Laboratory (INEEL). A brief epilogue draws the curtain on this book. A distinguishing feature of this book is the survey on the literature arranged section-wise at the end of each chapter for the convenience of the reader and a final bibliography at the end of the book for general reference on this research topic.

Desineni Subbaram Naidu
Pocatello, Idaho

Selahattin Ozcelik
Kingsville, Texas

Kevin L. Moore
Logan, Utah

April 2003

Acknowledgments

We would like to take this opportunity to express our sincere thanks to Dr.Herschell B.Smartt of the Idaho National Engineering and Environmental Laboratory (INEEL) in Idaho Falls, Idaho for his encouragement of this research activity. We particularly acknowledge his financial support, administered through Associated Western Universities (AWU), Salt Lake City, Utah, during the summer of 1997. We also acknowledge the support of INEEL through the University Research Consortium-funded project, *Advanced Control Welding Technology*, which stimulated the interest and need for the present research survey. We would also like to thank the entire URC project team, including Lyndon Brown, Bob Yender, Justin Tyler, Martin Murillo, Hardev Singh, Anna Mathews, and Mohamed Abdelrahaman. We also thank the College of Engineering at Idaho State University (ISU), which provided facilities for the URC project and for the preparation of this mongoraph. A special word of appreciation is also extended to our technical editor, Ms. Leonora Schaelling.

Due to the nature of this survey-oriented project, we depended a lot on the services of ISU Eli M. Oboler Library. In particular, the interlibrary loan staff, Nancy and Joan, deserve special mention for their ungrudging cooperation and timely action in bringing several of the copies of the references needed for the book and for their additional enthusiasm in getting the whole volume when only one particular article was requested! Also, our secretarial assistant Ann for doing lot of copying.

For preparing the final copy of this book, we used our favourite LaTeX format in PCTEX32[1] Version 4.0. The figures were drawn mainly using CorelDRAW[2] and exported into LaTeXdocument.

Finally, this survey-oriented book project would have taken a lot more time on our part, but for Subbaram's wife Sita, who had lot of patience in typing the huge bibliography files used in this survey. Thanks Sita! Finally, we enjoyed every moment of doing this project!

Desineni Subbaram Naidu
Pocatello, Idaho

Selahattin Ozcelik
Kingsville, Texas

Kevin L. Moore
Logan, Utah

---

[1] PCTEX32 and LaTeXare trademarks of Personal TEX, Inc., Mill Valley, CA.
[2] CorelDRAW is a registered trademarks of Corel Corporation.

Contents

Dedication	v
Preface	vii
Acknowledgements	ix
List of Figures	xvii

1 Introduction   1
    1.1 Introduction 1
    1.2 The Survey 4
    References List for Chapter 1 7

2 Gas Metal Arc Welding: Modeling   9
    2.1 Gas Metal Arc Welding 9
       2.1.1 Principles of Operation 11
       2.1.2 Arc Voltage and Self-Regulation of the Arc . . . 11
    2.2 Physics of Welding 14
       2.2.1 Physics of Arc 14
       2.2.2 Heat Transfer or Flow 15
       2.2.3 Other Works on Heat Flow or Transfer 17
       2.2.4 Cooling and/or Solidification Rates 20
       2.2.5 Arc Characteristics 20
    2.3 Melting Rate 22
    2.4 Metal Transfer Characteristics 23
       2.4.1 Globular Transfer 24
       2.4.2 Spray Transfer 25
       2.4.3 Streaming Transfer 25
       2.4.4 Short-Circuiting Transfer 25
       2.4.5 Pulsed Current Transfer 26
       2.4.6 Other Works on Metal Transfer 27
       2.4.7 Metal Transfer Experiments 28
       2.4.8 Physics of Metal Transfer 29
    2.5 Weld Pool 31
       2.5.1 Weld Pool and Weld Bead Geometry 32

	2.5.2	Other Works on Weld Pool	33
2.6	Process Voltages		35
	2.6.1	Cathode and Anode Voltages	36
	2.6.2	Arc Column Voltage	36
	2.6.3	Stick-Out Voltage	36
	2.6.4	Contact Tip Voltage	37
2.7	Heat and Mass Transfer		37
	2.7.1	Model for Heat and Mass Transfer	37
2.8	Process Variables		38
	2.8.1	Welding Current	39
	2.8.2	Polarity	39
	2.8.3	Arc Voltage (Arc Length)	40
	2.8.4	Travel Speed	40
	2.8.5	Electrode Extension (Stick-Out)	40
	2.8.6	Electrode Orientation	41
	2.8.7	Electrode Size	41
	2.8.8	Shielding Gases	41
	2.8.9	Classification of Process Parameters	42
2.9	INEEL/ISU Model		43
	2.9.1	Nomenclature	44
	2.9.2	Forces Affecting Droplet Dynamics	46
	2.9.3	Droplet Dynamics	47
	2.9.4	Model Equations	48
	2.9.5	Model Simplification and Linearization	50
2.10	Empirical and Statistical Models		52
2.11	Modeling by System Identification and Estimation		53
2.12	Intelligent Modeling		53
	2.12.1	Other Works on Intelligent Modeling	54
2.13	Other Issues on Modeling		55
	2.13.1	Dawn of GMAW	55
	2.13.2	Cost of GMAW	56
	2.13.3	Other Works on Modeling	56
2.14	Power Supplies		61
	2.14.1	Constant Current (CC)	63
	2.14.2	Constant Voltage (CV)	64
	2.14.3	Combined CC and CV Power Source	64

CONTENTS   xiii

        2.14.4 Pulsed Current 64
        2.14.5 Inverters 65
  2.15 Other Issues on Power Supplies 65
  2.16 Classification of References by Section 66
  References List for Chapter 2 67

3 Gas Metal Arc Welding: Sensing   95
  3.1 Classification of Sensors 95
  3.2 Conventional Method 99
  3.3 Computer-Based Measurements 99
  3.4 Welding Parameters Monitoring 100
        3.4.1 Temperature 100
        3.4.2 Welding Current 101
  3.5 Sensors for Line Following/Seam Tracking 101
  3.6 Arc Length Sensors 103
        3.6.1 Voltage Measurement 103
        3.6.2 Sound Measurement 103
        3.6.3 Laser (Range) Finders 103
        3.6.4 Light and Spectral Radiation Sensors 104
        3.6.5 Other Works in Arc Length Sensors 105
  3.7 Sensors for Weld Penetration Control 105
        3.7.1 Back-Face Sensing 106
        3.7.2 Front- or Top-Face Sensing 106
        3.7.3 Weld Pool Oscillation 110
        3.7.4 Droplet Transfer Frequency 111
  3.8 Sensors for Weld Pool Geometry 111
  3.9 Optical Sensors 113
  3.10 Sensors for Quality Control 114
  3.11 Intelligent Sensing 114
  3.12 Other Issues on Sensing 115
  3.13 Classification of References by Section 118
  References List for Chapter 3 121

4 GMAW: Automatic Control   147
  4.1 Automatic Welding 147
  4.2 Control of Process Variables 150
        4.2.1 Arc Length Control 150
        4.2.2 Control of Mass and Heat Transfer 151

		4.2.3	Control of Weld Temperature and/or Cooling Rate	152
		4.2.4	Control of Weld Pool and its Geometry	153
		4.2.5	Other Works on Control of Weld Pool Geometry	154
		4.2.6	Control of Droplet Transfer Frequency	154
		4.2.7	Control of Weld Penetration	156
		4.2.8	Control of Joint Profile (Fill Rate) and Trajectory	157
		4.2.9	Control of Other Variables or Conditions	159
	4.3	Classical Control: PI, PID and Others		160
	4.4	Multivariable Control		161
	4.5	Optimization and Optimal Control		163
	4.6	Adaptive Control		164
		4.6.1	ISU Adaptive Control Scheme	169
		4.6.2	Other Works on Adaptive Control	171
	4.7	Intelligent Control		172
		4.7.1	Fuzzy Logic	172
		4.7.2	Neural Networks and Fuzzy Logic	173
		4.7.3	Knowledge-Based and/or Expert System	175
		4.7.4	Other Works on Intelligent Control	177
	4.8	Statistical Process Control and Quality Control		178
	4.9	Other Control Methodologies and Issues		179
		4.9.1	Iterative Learning Control	179
		4.9.2	Feedback Linearization	180
		4.9.3	Relative Gain Array	180
		4.9.4	Other Works on Control	180
	4.10	Safety and Environmental Issues		183
	4.11	Classification of References by Section		184
	References List for Chapter 4			187
5	**Control of GMAW: A Case Study**			**219**
	5.1	Introduction		219
	5.2	Empirical Modeling of a GMAW Process		222
		5.2.1	Calibration of a GMAW Process	223
		5.2.2	Empirical Transfer Function Model	234
	5.3	SISO Current Control Using PI Controller		235
	5.4	Multi-Loop Control of the GMAW Process		239
		5.4.1	Relative Gain Array Analysis	240
		5.4.2	Multi-loop Control Experimental Results	240
		5.4.3	Disturbance Rejection Test	242

CONTENTS

- 5.5 Adaptive Control of GMAW Process 244
  - 5.5.1 Overview 245
  - 5.5.2 Model Simplification and Linearization 248
  - 5.5.3 Model for Heat and Mass Transfer 251
  - 5.5.4 Feasibility Region in the $\mathbf{G}-\mathbf{H}$ Plane 251
  - 5.5.5 Stability Analysis and Characterization of $\mathbf{G}-\mathbf{H}$ Plane 256
- 5.6 Control Strategy 257
  - 5.6.1 Formulation of the DMRAC 257
  - 5.6.2 Implementation of the DMRAC 260
  - 5.6.3 Experimental Results 262
- 5.7 Summary 268
- 5.8 Classification of References by Section 270
- References List for Chapter 5 271

6 Conclusions 275
- 6.1 Control Technology and Automation in Welding 275
- 6.2 Main Issues and Outlook 276
  - 6.2.1 Welding in Space Research 277
  - 6.2.2 Smart Robotic Welders and Manufacturing ... 277
- 6.3 Classification of References by Section 277
- References List for Chapter 6 279

Bibliography 281

Index 347

List of Figures

2.1	Simple diagram of the gas metal arc welding process.	12
2.2	Terminology of gas metal arc welding.	13
2.3	Self-regulation of arc voltage.	14
2.4	Transfer modes as a function of current.	24
2.5	Pulsing current for metal transfer.	26
2.6	Process voltage in the GMAW process.	35
2.7	Input and output variables of the welding process.	42
2.8	Relationship between welding parameters of the GMAW process.	43
2.9	Schematic diagram of the GMAW process.	44
2.10	Neural network estimator for weld pool geometry.	55
2.11	Classification of welding power sources.	62
2.12	Typical volt-ampere curves for: (a) constant current power sources, and (b) constant voltage power sources.	63
2.13	Typical volt-ampere curves for a combined CC and CV power source.	65
3.1	Principle of weld pool oscillations.	98
3.2	Principle of computer-based measurements.	100
3.3	Schematic of laser shadow motion sensing method.	107
3.4	Simplified schematic of ultrasonic sensing method	107
3.5	Simplified schematic of phase locked loop method for pulsing	110
4.1	(a) Open-loop and (b) Closed-loop control systems.	149
4.2	Feedforward control system.	150
4.3	PID control system for arc length regulation.	151
4.4	Frequency Modulated Pulse Current Feedback System	152

4.5	Schematic for a weld pool geometry (width and depth) control system.	153
4.6	PID control system for droplet frequency regulation.	155
4.7	Welding equipment and instrumentation for experimental facility.	156
4.8	Intelligent control of the GMAW process.	158
4.9	An early feedback control system for a welding process.	160
4.10	PI control of the GMAW process.	161
4.11	A multivariable feedback control system for the GMAW process.	162
4.12	A multivariable linearized feedback controller for the GMAW process.	163
4.13	Basic principle of adaptive control.	165
4.14	Alternative scheme of adaptive control.	166
4.15	MRAC for plate temperature.	167
4.16	Adaptive thermal control system.	168
4.17	Adaptive thermal control system using a Smith predictor.	169
4.18	Direct model reference adaptive control of the GMAW process.	170
4.19	Self-learning fuzzy neural control system for arc welding processes.	174
4.20	Experimental setup for neurofuzzy model-based control.	175
4.21	Artificial intelligence system for a welding process.	176
5.1	The "big-picture" of GMAW process control.	221
5.2	Experimental data - contour plot of average current (amps).	226
5.3	Data - contour plot of average arc voltage (volts).	227
5.4	Simulation data - average current from the fifth-order model for the whole data set.	228
5.5	Simulation data - best squared error for the whole data set applications.	229
5.6	Simulation data - average current from the fifth-order model for individual data sets.	230
5.7	Simulation data - best squared error for individual data sets.	230
5.8	Simulation data - best R_a parameter for individual data sets.	231

LIST OF FIGURES

5.9 Simulation data - best V_o parameter for individual data sets. ... 231
5.10 Simulation data - best E_a parameter for individual data sets. ... 232
5.11 Simulation data - best C_1 parameter for individual data sets. ... 232
5.12 Simulation data - best C_2 parameter for individual data sets. ... 233
5.13 Current response to a step increase in the wire-feed speed. 234
5.14 Voltage response to a step increase in the open-circuit voltage. ... 235
5.15 Current response to a step increase in the open-circuit voltage. ... 236
5.16 Voltage response to a step increase in the wire-feed speed. 237
5.17 Closed-loop system with PI controller. ... 237
5.18 Current response, $K_p = 0.5$, $K_i = 5$, desired current=260A, actual current=260.0033A for $3 > t \leq 5$ and desired current=240A, actual current=240.0955A for $5 > t \leq 7$. ... 238
5.19 Control signal, open-circuit voltage, for $K_p = 0.5$, $K_i = 5$, desired current=260A for $3 > t \leq 5$ and desired current=240A for $5 > t \leq 7$. ... 238
5.20 Arc voltage, for $K_p = 0.5$, $K_i = 5$, desired current=260A for $3 > t \leq 5$ and desired current=240A for $5 > t \leq 7$. ... 239
5.21 Multi-loop control of the GMAW process. ... 240
5.22 Experimental results: current response, with simulation data. ... 242
5.23 Experimental results: voltage response, with simulation data. ... 243
5.24 Experimental results: open-loop current response, step disturbance. ... 244
5.25 Experimental results: open-loop voltage response, step disturbance. ... 245
5.26 Experimental results: closed-loop current response, disturbance rejection. ... 246
5.27 Experimental results: closed-loop wire-feed speed response, disturbance rejection. ... 247

5.28 Experimental results: closed-loop voltage response, disturbance rejection. 253
5.29 Experimental results: open-loop open-circuit voltage response, disturbance rejection. 253
5.30 Feasible region in the $G - H$ plane. 254
5.31 Weld speed with respect to $G - H$ values. 254
5.32 Current changes in the $G - H$ plane. 255
5.33 Distribution of the system response in the $G - H$ plane – (+): underdamped, (o): overdamped. 257
5.34 Eigenvalue distribution in the $G - H$ plane. 258
5.35 Closed-loop GMAW with DMRAC. 263
5.36 Current response, desired current = 260A, actual current = 260.3824A. 265
5.37 Arc voltage response, desired arc voltage = 29V, actual arc voltage = 28.6023V. 265
5.38 Current response, desired current = 240A, actual current = 240.9829A. 266
5.39 Arc voltage response, desired arc voltage = 24V, actual arc voltage = 24.8629V. 266
5.40 Current response, desired current = 260A, actual current = 261.9960A for $3 < t \leq 5$, desired current=240A, actual current=241.5589A for $5 < t \leq 7$. 267
5.41 Arc voltage response, desired arc voltage = 29V, actual arc voltage = 28.0629V for $3 < t \leq 5$, desired arc voltage=24V, actual arc voltage=24.9090V for $5 < t \leq 7$. . . 267

Chapter 1

Introduction

1.1 Introduction

Manufacturing processes *Joining* and *cutting* are very important for any industrial activity. Of these, cutting is a relatively straightforward and well-known process, which may be done mechanically or thermally, using thermal sources such as oxygen or a plasma[1]. On the other hand, the process of joining is more involved. The basic joining processes are [2]

1. mechanical joining,

2. adhesive bonding,

3. brazing and soldering, and

4. welding.

The focus of this monograph is on the fourth joining process, welding, and specifically Gas Metal Arc Welding (GMAW). According to American Welding Society [1], welding is defined as *a localized coalescence of metals or nonmetals produced either by heating the materials to the welding temperature, with or without the application of pressure, or by the application of pressure alone, with or without the use of filler metal.* Although there are some 40 or so welding processes, only a few processes are important. In welding, the shielded metal arc welding (SMAW) and GMAW are some of the most widely used in industry.

The term *arc welding* refers to a broad group of welding processes that employ an electric arc as the source of heat to melt and join metals. It is believed that in the entire metal fabrication industry, arc welding is the third largest job category behind assembly and machining in metal fabrication industry [3].

The physics of welding deals with the complex physical phenomena associated with welding, including electricity, heat, magnetism, light, and sound. In particular, many welding processes require the application of heat or pressure, or both, to produce a suitable bond between the parts being joined. A common means of heating for welding is by the flow of current through electrical contact resistance at the joining surfaces of two work pieces. Welding processes that acquire heat from external sources are usually categorized with the type of heat source used. The processes in this category, with a brief definition, are given below [4].

1. Arc Welding

    (a) *Shielded Metal Arc Welding (SMAW)* is an arc welding process where an electric arc that is maintained between the tip of a covered electrode and the surface of the base metal produces the heat required for joining[5].

    (b) *Gas Tungsten Arc Welding (GTAW)* is an arc welding process that uses the arc between a non-consumable tungsten electrode and the weld pool with a shielding gas.

    (c) *Gas Metal Arc Welding (GMAW)* is an arc welding process that uses an arc between a consumable electrode and the welding pool with a shielding from externally supplied gas without any application of pressure. In Europe, GMAW is also called metal inert gas (MIG) or metal active gas (MAG) welding [2].

    (d) *Flux-Cored Arc Welding (FCAW)* is an arc welding process that uses an arc between the consumable electrode and the weld pool with a shielding from a flux contained within the tubular electrode with or without additional shielding from an externally supplied gas.

    (e) *Submerged Arc Welding (SAW)* is an arc welding process that produces joining of metals by heating them with an arc between a metal electrode and the work piece. The arc and the molten metal are "submerged" in a flux on the work piece.

(f) *Electro-gas Welding (EGW)* is an arc welding process that uses an arc between a consumable electrode and the weld pool, using auxiliary gas shielding around a flux-cored electrode.

(g) *Plasma Arc Welding (PAW)* is an arc welding process that produces coalescence of metals by heating them with an arc between an electrode and the workpiece with shielding from an ionized gas.

2. *Resistance Welding (RW)* is a process in which the heat required for welding is produced by the resistance to the flow of electric current passing through the parts to be welded.

3. *Electro-Slag Welding (ESW)* is a welding process that produces joining of metals with molten slag that melts the filler metal and the surfaces of the workpieces to be welded with the shielding being provided by the slag.

4. *Oxyfuel Gas Welding (OFW)* is a welding process that uses combustion with oxygen as a heating medium by mixing fuel gas and oxygen inside a mixing chamber.

5. *Thermit Welding (TW)* is a welding process that heats the metals to be welded with super heated molten metal from an aluminothermic reaction between a metal oxide and aluminum.

6. *Diffusion Welding (DFW)* is a solid-state welding process using the pressure at elevated temperatures with no microscopic deformation or relative motion of the workpieces.

7. *Electron Beam Welding (EBW)* is a welding process that uses a concentrated beam of high-velocity electrons to provide a heating source.

8. *Laser Beam Welding (LBW)* is a welding process that uses a focused high-power monochrome light to produce a deep penetration column of vapor to the base metal.

9. *Friction Welding (FRW)* is a solid-state welding process that uses compressive force to contact workpieces rotating or moving relative to one another to produce heat.

10. *Ultrasonic Welding (USW)* is a solid-state welding process that produces a weld by local application of high frequency vibratory energy.

11. *Spot, Seam, and Projection Welding* are resistance welding processes in which joining of metals is produced by the heat generated by the resistance of the work to the passage of electric current.

12. *Flash, Upset, and Percussion Welding* are a group of welding processes used to join parts by simultaneously making a weld across the joint area without using filler metal.

The practice of industrial welding is heavily dependent on the knowledge and vast experience of the welder and as such, it is at present more an art than a science. However, in order to make the welding process more *automated* and less human (welder) dependent, in the last two decades, significant efforts have been made to introduce the ideas of feedback in order to *control* the welding process to achieve a *good* weld. Modeling, sensing and control techniques and strategies are the subject of this monograph.

1.2 The Survey

Although, the subject presented here was originally intended to be focused on automatic control techniques for GMAW, the authors soon found that it was necessary to introduce other aspects such as modeling, power supplies, and sensing. Thus, in this research survey, efforts have been directed towards providing a comprehensive review of the GMAW process in general and the modeling, sensing, and automatic control aspects in particular.

The topic of arc welding has a vast research literature. However, the literature on modeling, sensing, and control of the GMAW process is not as extensive, and there are only a few surveys on the topic. Indeed, a very good general presentation on the need for modeling and control of manufacturing processes in general and welding in particular

1.2. THE SURVEY

was most revealing [6]. The author reviewed two decades (1970-1990) of manufacturing control research in the ASME Journal of Dynamic Systems, Measurement and Control and found that there are only 25 articles published in the Journal in the area of manufacturing and out of these, there were only six papers on arc welding! More references can be found in three recent literature survey articles, including one on *expert robotic* welding systems with 92 references [7], and another on sensory feedback control for *robotic* arc welding, with 83 references [8]. While these surveys consider only *robotic* welding, a third tutorial type of survey presented during a keynote address by G.E. Cook, et al. can be found in [9]. In particular, the survey by Cook, et al. (with 118 references) focused on sensing, modeling and control of welding processes. The survey presented here also focuses on various aspects of modeling, sensing, and automatic control of the GMAW process, but it considers a significantly expanded number of references (over 600!). The main purpose of this research monograph is to provide researchers with the status of the state-of-the-art of the work in the areas of modeling, sensing, and control of the GMAW welding process. Occasionally, the gas tungsten arc welding (GTAW) process is also touched upon.

Besides the review articles indicated above, the interested reader should also refer to several excellent welding reference books, including Volume 1 [1] and Volume 2 [4] for all aspects of welding, the classic text for modeling [10], and a recent (and the only) book for sensing and control in arc welding (although the material is exclusively based on the Japanese works) [11]. From the historical perspective, the reader is also highly recommended to look at the first survey on arc welding for an extensive literature survey (with 178 references) on the arc physics and metal transfer up to February 1942 [12]. This survey was prepared under the auspices of the Literature Division of the Engineering Foundation Welding Research Committee and was done in two parts, one connected with low-amperage arcs and the second part related to welding arcs.

The subject matter is organized topically starting first with modeling, then considering sensing, and then finally control. Also included is a case study highlighting some experimental work done by the authors at Idaho State University. In each section we describe the topic and summarize results from the literature. At the end of each section a table is given that gives reference numbers by topic. References are listed at the end of each chapter in the order they were discussed and are also collected at the end of the book in alphabetical order. It should be noted that references may be discussed more than once. For example, if we talk about a particular article on adaptive control systems for controlling the weld geometry, then that particular article is discussed both under the category of adaptive control and the category of control of weld geometry. We should also point out that the monograph focuses primarily on modeling, sensing, and control of gas metal arc welding. The topic robotic arc welding, which in itself is a major research area, is not covered in this monograph.

Finally, the authors wish to stress that they have tried their best to compile all the relevant references on this topic. However, it is inevitable that some citations have been overlooked. In case some references that should have been included are missing from the survey, it is purely unintentional. The authors would greatly appreciate having such omissions brought to their attention.

Table 1.1: Section by Section List of References

Section	Reference Numbers
1.1 Introduction	[1]-[7]
1.2 Survey	[6]-[12]

References List for Chapter 1

[1] L. P. Connor. Welding Handbook. Eighth Edition. volume 1. Miami, FL, American Welding Society, 1987.

[2] J. Norrish. Advanced welding processes. Institute of Physics Publishing. Bristol, UK, 1992.

[3] J. G. Holmes and Resnick. A flexible robot arc welding system. Proceedings of the American Society of Mechanical Engineers, Pages 1-15. 1979.

[4] R. L. O'Brien. Welding handbook: Welding processes. 2, Eighth Edition. Miami, FL, American Welding Society, 1991.

[5] T. W. Eagar. Resistance welding: a fast, inexpensive and deceptively simple process. Proceedings of the 3rd I. Conference on Trends in Welding Research. Editor, S. A. David and J. A. Vitek. Gatlinburg, TN. Pages: 347-351, June, 1992.

[6] D. E. Hardt. Modeling and control of manufacturing processes: getting more involved. Transactions of the ASME, Journal of Dynamic Systems, Measurement and Control, 115, 291-300. June, 1993.

[7] P. Sicard and M.D. Levine. An approach to an expert robot welding system. IEEE Transactions on Systems, Man and Cybernetics. 18(2), Pages 204-222. March/April,1988.

[8] G. E. Cook. Through-the-arc sensing for arc welding. Proceedings of the 10th NSF Conference on Production. Research and Technology, Pages 141-151. February28-March2, 1983.

[9] G. C. Cook, K. Anderson and R. J. Barrett. Feedback and adaptive control in welding. Proceedings of the 2nd International Conference on Trends in Welding Research, Gatlinburg, TN, May. S. A. David and J. M. Vitek, Key Note Address. Pages 891-903, 1989

[10] J. F. Lancaster. The Physics of Welding. Pergamon Press, Second. Oxford, UK, 1986.

[11] H. Nomura. Sensors and Control Systems in Arc Welding. Chapman & Hall, London, UK. English Translation of the Original 1991 Japanese Edition, 1994.

[12] W. Spraragen and B. A. Lengyel. Physics of the arc and the transfer of metal in arc welding: A review of the literature to February 1942", Welding Journal,Volume 1,January, Pages 2s–42s, 1943

Chapter 2

Gas Metal Arc Welding: Modeling

In this chapter, aspects of both modeling and power supplies are presented. The modeling of a welding process in general means the derivation of a set of mathematical equations (ordinary differential equations for lumped parameter systems, partial differential equations for distributed parameter systems) describing the physical process by means of fundamental principles of science or statistical and/or experimental techniques. It is our perspective that effective feedback controller design cannot be accomplished without a framework within which it is possible to describe and understand the behavior of the system to be controlled. Thus, it is important to have a model. In this chapter we describe the literature on modeling the Gas Metal Arc Welding (GMAW) process. We begin with a description of GMAW. Then we discuss various aspects of the physics of welding. The chapter concludes with the presentation of a specific model developed by the authors and their colleagues at a U.S. Department of Energy national laboratory, the Idaho National Engineering and Environmental Laboratory (INEEL).

2.1 Gas Metal Arc Welding

The welding process is a multi-energy process involving such different phenomena as plasma physics, heat flow, fluid flow, and heat and metal transfer, etc [13]. The basic concept of the GMAW process was known by the 1920s, but only in 1948 was the process made commercially available [14]. At first, the process included an inert gas for shielding and hence was called *Metal Inert Gas Welding (MIGW)*. Subsequent developments used reactive gases such as CO_2 and other gas mixtures, which led to the terminology *Gas Metal Arc Welding (GMAW)*, which implied using both inert and reactive gases.

The reason for the acceptance of the GMAW process for almost all the industrial applications is due to its versatility and specific advantages, such as those listed below[14]:

1. GMAW is the only consumable electrode welding process that can be used for welding all commercial metals and alloys.

2. GMAW welding can be done in all positions, unlike in submerged metal arc welding.

3. Because of the continuous electrode feed, the metal deposition rates in GMAW are significantly higher than shielded metal arc welding (SMAW).

4. Due to higher metal filler deposition rates, welding speeds in GMAW can be higher than those obtained with SMAW.

5. Because the wire feed is continuous with GMAW, longer welds can be done without stops and starts.

6. GMAW has no restriction on the length of the electrode as in SMAW.

7. With spray transfer in GMAW, deeper penetration of the weld is possible compared to SMAW.

8. Due to the absence of slag, there is less problem with cleaning.

However, there are some limitations of the GMAW process:

1. The welding equipment is more complex and hence costlier and less portable compared to SMAW.

2. Protection against air drafts is required.

3. Higher levels of radiated heat and arc intensity are produced.

2.1. GAS METAL ARC WELDING

The GMAW process can be used to join virtually any metal using many joint configurations, and in all welding positions [15]. The welding process in general is a very complex process due to the fact that the process involves many scientific and engineering disciplines such as chemistry, physics, metallurgy, materials science, and mechanics. The process also involves a complex interaction of solid, liquid, gaseous, and plasma-state phenomena. The complexity of the process is due to the fact that a large number of these phenomena takes place *simultaneously* in a relatively small volume (0.1 to 10 mm^3) over a short distance (1 to 20 mm), and frequently over short periods of time [16].

The GMAW process is relatively a complex and "dirty" process, but often the most widely-used process in industry, particularly in automatic robotic arc welding. In real life and practical applications, the GMAW process exceeds by over 10 fold the "clean" GTAW process [17].

2.1.1 Principles of Operation

Gas metal arc welding (GMAW) is a welding process where the heat is generated by an electric arc incorporating a continuous-feed consumable electrode that is shielded by an externally supplied gas. A simple schematic diagram of the GMAW process is shown in Figure 2.1[14]. Figure 2.2 shows the terminology used with the GMAW process.

Besides the welding gun, the actual equipment required for the GMAW process includes an electric power supply, the electrode wire-feed unit, and a source of shielding gas. The gun guides the electrode wire, current wire, and shielding gas tube. As described in the next section, self-regulation of the arc length is maintained by a constant voltage power supply with a constant wire-feed speed unit. Alternatively, a constant-current voltage supply can be used with arc voltage controlling the wire-feed speed.

2.1.2 Arc Voltage and Self-Regulation of the Arc

The voltage drop across the arc (arc voltage) is directly proportional to arc length. Hence, the arc voltage is controlled by changing the arc length. There is a minimum voltage for striking the arc, and the open-circuit voltage of the power supply is obviously higher than the

CHAPTER 2. GAS METAL ARC WELDING: MODELING

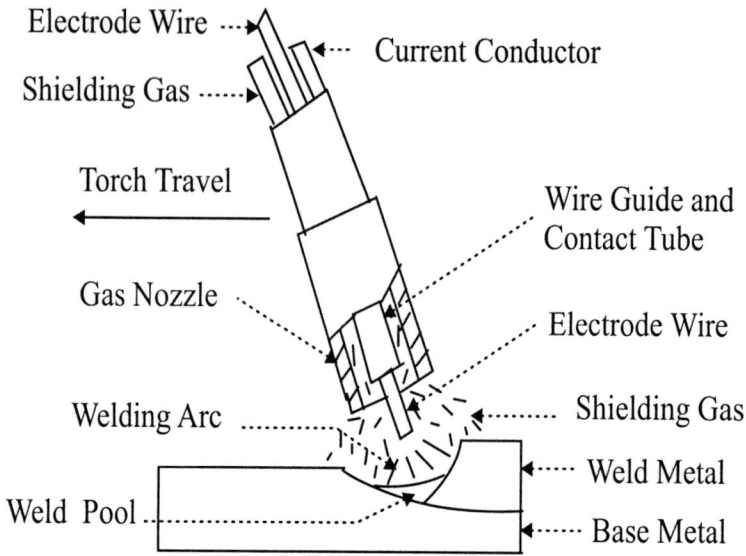

Figure 2.1: Simple diagram of the gas metal arc welding process.

arc-striking voltage.

In order to understand the mechanism of maintaining the arc, consider the constant-voltage (CV) and constant-current (CC) volt-ampere characteristics of a typical power supply along with arc voltage curves as shown in Figure 2.3 [18]. Notice that the change in current for a change in arc length (from high to low) for a CC power supply is much less than that for CV power supply. The intersection of an arc voltage curve with the voltage source curve is called the *operating point* for the power supply. The operating point may change continuously during welding operation.

1. For the case of a *CV power supply* with a constant wire-feed speed, consider the operating point P. A small decrease in arc length and hence a small decrease in arc voltage results in a relatively large increase in welding current. This large increase in current increases the melt-off rate, thereby increasing the arc length and hence increasing the arc voltage back to its normal operating point. The opposite action happens when the arc length is increased from any operating point P. Thus, the GMAW process

2.1. GAS METAL ARC WELDING

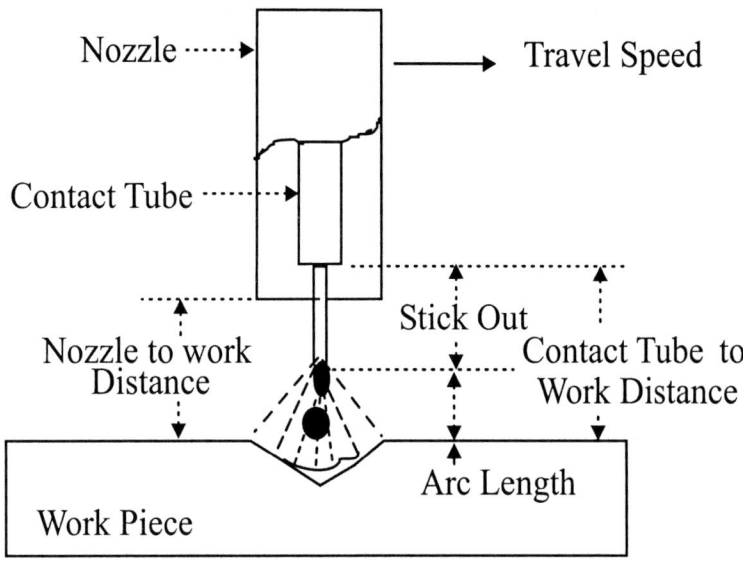

Figure 2.2: Terminology of gas metal arc welding.

employing a CV power supply with fixed wire-feed speed has a *self-regulating* feature, which is very good for a semi-automatic welding system. Thus, while using the combination of a constant voltage/constant wire-feed speed unit, a change in torch position causes a change in arc current that exactly compensates the electrode stick-out (electrode extension), thereby maintaining a constant arc voltage.

2. For the case of a *CC power supply* with a constant wire-feed speed, a similar sequence of events takes place, except the change in current is much less, and thus it can take more time to self-regulate the system. In a truly CC (steeply dropping) system, the change in current is much less, thereby meaning that the welder (by changing the arc length) has little or no control on the current. Also, with a CC power supply with a steep dropping characteristic, the arc will maintain a fixed length only if the contact-tube-to-workpiece distance remains constant with a constant wire-feed rate. However, in practice, since this distance will change, the arc will then either tend to *burn back* to the con-

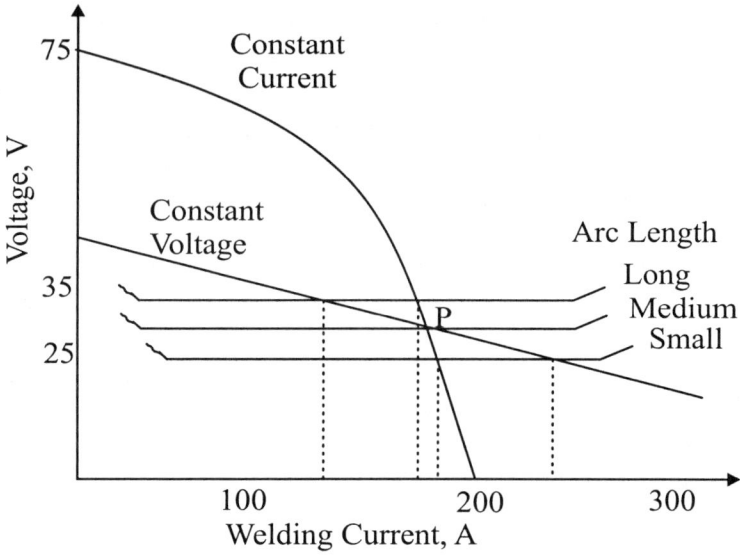

Figure 2.3: Self-regulation of arc voltage.

tact tube or *stub* into the workpiece. This can be rectified by using a voltage-controlled electrode-feed control system. If the arc voltage increases (decreases), the motor speeds up (down) to hold the arc length constant.

A study on the dynamic characteristics of self-regulation of the arc in a GMAW process is given in [19]. An analytical model using a classical frequency response method was developed to determine the degree of arc length regulation as a function of the frequency of torch-to-workpiece (also called contact-tube-to-workpiece) distance change. It was found and experimentally verified that the regulation of arc length decreases with increasing frequency.

2.2 Physics of Welding

2.2.1 Physics of Arc

Arc is the main part of the physics of the welding process. In welding, a high energy density heat source is applied to the parts or surfaces

2.2. PHYSICS OF WELDING

to be joined and is moved along the path of the intended joint. When a filler may be added, the heat source must also be sufficient to melt the filler material as it is delivered to the joint surface. In-depth treatments of arc modeling can be found in [20, 21, 22, 23]. The complete understanding of the definition of the arc, the consumable-electrode welding arc, and the function of the arc are given in the series of classic contributions [24, 25, 26]. The different arc types as well as the relevant welding parameters are discussed in [27]. Another study in [28] investigates the relationship between the arc light intensity and arc length through analytic modeling. Using the heat balance in the plasma, the arc light intensity is derived as a function of the arc length and the welding current. Computer simulation of arc behavior and bead appearance for a GMAW process is developed in [29]. Arc initiation with GMAW depends on various factors such as wire speed, wire-to-base metal contact resistance, geometry, etc. Farson [30] has studied arc initiation in GMAW.

2.2.2 Heat Transfer or Flow

In the case of arc welding, the energy input is usually the arc energy input. Arc energy input is the quantity of energy transferred per unit length of weld from a traveling heat source expressed in joules per meter (m) or millimeter (mm). Thus, the energy input (H) is defined as the ratio of the total input power (P) of the heat source in watts to its travel velocity (R) expressed in mm/sec, or

$$H = P/R \qquad (2.1)$$

If the heat source is an arc, then the heat input energy (to a first approximation) is

$$H = EI/R \qquad (2.2)$$

where E is the voltage (volts) and I is the current (amps). Taking efficiency (η_t) of the heat transfer into account, the net energy input (H_{net}) is given as

$$H_{net} = \eta_t H = \frac{\eta_t P}{R} = \frac{\eta_t EI}{R} \qquad (2.3)$$

Typically, the efficiency factor has a high value of between 0.66 and 0.85 for the GMAW process and between 0.21 and 0.48 for the GTAW process [31].

In heat transfer, we further note that the melting efficiency is the fraction of the net energy input (H_{net}) used for melting the metal. Also, there is a specific theoretical quantity of heat Q required to melt a given volume of the metal from a cold start. This quantity Q, in Joules/mm^3, is

$$Q = \frac{(T_m + 273)}{300,000} \qquad (2.4)$$

where T_m is the melting temperature in degrees Celsius of the metal. Taking the melting efficiency η_m into account, we have

$$\eta_m = \frac{QA_w}{H_{net}} = \frac{QA_w v}{\eta_t P} = \frac{QA_w v}{\eta_t EI} \qquad (2.5)$$

where A_w is the cross-section of the weld metal, which can be written as the sum of the area of the base metal that was melted (A_m) and the area of the weld metal (also called area of reinforcement) (A_r). A_r also represents the filler metal added to the weld.

Using the relations (2.5) and (2.3), we get an important relation between the weld metal cross section A_w and the energy input H as

$$A_w = \frac{\eta_m H_{net}}{Q} = \frac{\eta_t \eta_m H}{Q} \qquad (2.6)$$

The transfer of heat energy in the weldment is governed by the conduction of heat, which is described by the following partial differential equation [15]

$$\frac{\partial}{\partial x}\left[k(T)\frac{\partial T}{\partial x}\right] + \frac{\partial}{\partial y}\left[k(T)\frac{\partial T}{\partial y}\right] + \frac{\partial}{\partial z}\left[k(T)\frac{\partial T}{\partial z}\right] = \rho C(T)\frac{\partial T}{\partial t} - Q \qquad (2.7)$$

where x is the coordinate in the welding direction (mm), y is the coordinate transverse to the weld (mm), z is the coordinate normal to weldment surface (mm), T is the temperature in the weldment, (^0C); $k(T)$ is the thermal conductivity of the metal ($J/mm.s.^0C$), ρ is the density of metal (g/mm^3), C is the specific heat of the metal ($J/g.^0C$), and Q is the rate of internal heat generation (W/mm^3).

2.2. PHYSICS OF WELDING

A complete theoretical analysis of heat distribution was done by Rosenthal [32], and this model was modified by Jhavari, et al. [33], taking into account the plate thickness and the heat loss due to radiation.

2.2.3 Other Works on Heat Flow or Transfer

An early development of a general set of charts showing the effect of plate thickness, thermal properties, and operating variables of the welding process on thermal properties such as cooling rate and peak temperature distribution can be found in [33].

In [34], it is stated that during heat transfer with GMAW and plasma-GMAW, the heat value of transferring metal drops appears to determine the total cross-sectional area of weld penetration while the impact of drops on the liquid metal weld pool determines the depth of penetration.

A study on the physical processes governing the generation and the flow of heat in a consumable mild steel wire to determine the influence of welding parameters on melting rate and drop temperature for a GMAW process can be found in [35].

Modeling of heat transfer and thermal behavior of metals during welding, with reference to the GTAW process, are discussed in [36, 37, 38].

In [39], a simple FORTRAN-based computer program has been developed for calculating the heat affected zone (HAZ) and cooling rate (CR) for rectangular plates.

The work [40] presents experimental investigations on drop detachment and drop velocity in a pulsed GMAW process. The investigators measured the velocity of drops detached and the diameter of the drop neck during the detachment process with the aid of a high-speed photography arrangement consisting of a laser, lenses, optical filter, and fixed and rotating mirrors. It is found that the pendent drop contracts when the pulse starts and becomes unstable when the diameter reaches a critical value.

An in-depth study of the effect of welding heat input and preheating on cooling rates can be found in [41].

Experiments to determine the values of welding parameters that affect heat input calculations for pulsed GMAW are given in [42].

A new research paradigm called the inverse problem, involving an interactive combination of complex experiments and analysis, is discussed in [43] for heat transfer with applications to welding and solidification.

A study on the three-dimensional analysis of heat and fluid flow in GMAW using boundary fitted coordinates is presented in [44].

An in-depth study on the metal transfer (heat and fluid flow), both analytically and experimentally, is given in [45]. The commentary on this work is given by Lesenewich [46].

Matsunawa [47] gives a keynote address that considers a brief history of studies on heat and mass transfer in arc welding, possible factors to induce flow in a puddle and their effects on penetration shape, some experimental verifications of the mathematical model, and future outlook in the field.

An axisymmetric thermo mechanical analysis of a stationary gas metal arc weld, employing sequential coupling of the thermal and stress analysis using finite element method (FEM) is given in [48].

In [49], a computer program was developed for the computation of thermal cycles at heat affected zones with GMAW of medium thickness plates based on the computation model for quasi-steady heat transfer problem and the boundary element method.

The development of a model of the GMAW process to predict the three-dimensional transient temperature distribution in the workpiece and numerical simulation using FEM are given in [50, 51]. The authors used pinch instability theory (PIT) and the static force balance theory (SFBT) of the drop detachment and obtained expressions for the various characteristics of the drop such as drop radius, drop velocity, and drop frequency.

In [52], Ushio introduces a three-dimensional model for heat and fluid flow in a moving gas metal arc weld pool. The model takes the mass, momentum, and heat transfer of filler metal droplets into consideration and quantitatively analyzes their effects on the weld bead shape and weld pool geometry.

Kim and Basu [53] have developed an unsteady two-dimensional (2D) axi-symmetric model for investigating the heat and fluid flows in weld pools. Based on that the weld bead geometry, and the velocity and temperature profiles for the GMAW process have been determined.

2.2. PHYSICS OF WELDING

In the mathematical formulation electromagnetic, buoyancy, surface tension, and drag forces are considered for weld pool convection.

In [54], calculation of the thermal cycles near the weld region during gas metal arc welding was studied. An estimation of the depth of weld penetration, the geometry of the weld pool and the cooling rates are calculated based on predicted thermal cycles.

A heat flux distribution model, suitable for larger surface deformation of a weldpool, of GMAW WAS developed in [55]. The spatial distribution of the heat flux has an effect on the weld-pool shape and solidification process, which in turn affects the structure and properties of the weldment.

Wu, in [55], developed a model for the distribution of the heat content of filler metal droplets inside the gas-metal arc weld pool. It is known that this distribution has an effect on the weld bead dimensions and the weld thermal cycle and is considered as an internal heat generation term.

In GMAW processes, the time-dependent temperature and displacement fields generated by the weldment are the parameters of interest. In [56], a measurement system that combines both full-field temperature and three-dimensional displacement measurement capabilities is presented.

A three-dimensional finite element simulation of the welding process is presented in [57]. The analysis involves de-coupled heat transfer and thermo-mechanical analysis for the determination of residual stress and distortion.

In [58], fluid flow and heat transfer during GMAW of HSLA-100 steel were studied using a transient, three dimensional, turbulent heat transfer and fluid flow model. The temperature and velocity fields, cooling rates, and shape and size of the fusion and HAZs were calculated. To better understand a weld metal microstructure, a continuous cooling transformation (CCT) diagram was computed.

Experiments have been done in [59] to simulate the effect of Marangoni convection on weld pool shape. Stationary welds $NaNO_3$ and Ga were made with a defocused CO_2 laser beam to simulate the effect of Marangoni convection on the shape of arc weld pools without a surface-active agent. It was proposed that in the absence of both a surface-active agent and a significant electromagnetic force, the pool bottom convexity increases with increasing Pe.

2.2.4 Cooling and/or Solidification Rates

The solidification time S_t, in *sec*, is given by

$$S_t = \frac{LH_{net}}{2\pi k\rho C(T_m - T_0)^2} \tag{2.8}$$

where L is the heat of fusion, J/mm^3, C, k, and ρ are constants, and T_m and T_0 are the metal temperature and the ambient temperature, in ^0C, respectively.

Dorschu [60] experimentally studied the dependence of cooling on GMAW process variables and found that the cooling rates are affected by the torch speed and pre-heating of the workpiece. Further studies on cooling and solidification include [36, 22, 23].

Garland [61] studied principles involving in solidification processes and the control of weld pool solidification to produce weld metals possessing enhanced properties.

Derivations for solidification of the molten metal in terms of the liquid phase angle and for the bead formation assuming a simple harmonic wave of the molten metal are given by [62].

2.2.5 Arc Characteristics

A welding arc can be viewed as a gaseous conductor that converts electrical energy into heat energy. The various characteristic features of the welding arc are the plasma, temperature, radiation, electrical features, magnetic fields, and arc blow. We briefly describe each of these (except radiation) in this section.

Plasma

The arc current is carried by a plasma, the ionized state of a gas, which is composed of nearly equal number of electrons and ions. The electrons, which support most of the current conduction, flow out of a negative (cathode) terminal and move towards a positive (anode) terminal. In the case of arc welding one of these cathode or anode electrodes serves as the workpiece and the other is called the *electrode*.

2.2. PHYSICS OF WELDING

Mixed with the plasma are the other things including molten metals, slags, vapors, and molecules. The formation of plasma is governed by the advanced concepts of the *Ideal Gas Law* and the *Law of Mass Action* [15]. The plasma typically has an axial temperature in the range of 5,000 to $15,000^0 K$ [63]. Craig [64] gives an excellent review of the plasma arc process and the work in [65] discusses the placement of the plasma arc sheath to achieve several productivity applications.

Temperature

The temperatures of the welding arc are in the range of 5,000 to $30,000^0$ K depending on the condition of the plasma and the current flowing through it. Some special arcs of extreme power may reach an axial temperature of $50,000^0$ K. In most cases, the temperature of the arc is measured by techniques using the spectral radiation emitted by the arc.

Electrical Features

A welding arc acts like an impedance to the flow of electric current in the power supply circuit. The electrical power dissipated in the arc is the sum of the power in the three regions of the arc: the anode, the cathode, and the plasma. Thus,

$$P = I(E_a + E_c + E_p) \tag{2.9}$$

where P is the power (W), I is the current (A), E_a is the anode voltage (V), E_c is the cathode voltage (V), and E_p is the plasma voltage (V).

Magnetic Fields

Magnetic fields, either induced or self, interact with the arc current to produce force fields that cause an arc deflection that is usually called *arc blow*. Further, the magnetic fields affect plasma streaming and metal transfer. The effects of magnetic fields on welding arcs are determined by the Lorentz equation, according to which the force produced is proportional to the vector cross product of the external field strength and the arc current.

Arc Blow

The phenomenon where the arc has a tendency to be forcibly directed away from the point of welding, is often termed *arc blow*. In general, the arc blow is the result of two conditions: (1) the change in current direction and (2) the presence of magnetic materials around the arc. Low arc voltage results in shorter, stiffer arcs that resist arc deflection better than a higher arc voltage. Physics and chemistry of welding processes in general can be found in [14, 66].

2.3 Melting Rate

As described earlier, the heat energy in an arc is generated by electrical reactions at the anode and cathode regions within the plasma. Portions of this energy will melt the electrode. The melting rate is primarily affected by the current or cathode heating. The melting rate MR is given by [67]

$$MR = aI + \frac{\beta}{a_w} l_s I^2 \qquad (2.10)$$

where a and β are constants, l_s electrode resistivity, I is the welding current and a_w is cross sectional area of the wire. For further discussion on melting rate, see the classic paper by Lesnewich [68].

Chandel [69] has studied the effects of welding variables and parameters such as welding current, arc voltage, wire diameter, electrode extension, electrode polarity, and power source type on the melting rates for submerged arc welding process. The calculation of the ohmic heating in the electrode extension of different filler wires and a simple relation for the melting rate are given in [70]. Tusek [71] gives a mathematical model for the calculation of melting rates obtained in gas-shielded arc welding with a multiple-wire electrode and in submerged arc welding with a multiple-wire electrode. Bingul [72] gives a model correlating the anode temperature profile with the dynamic melting rate in GMAW.

2.4 Metal Transfer Characteristics

In general, the consumable electrode arc welding processes are preferred because the filler metal is deposited more efficiently at higher rates of deposition than with other welding processes [14]. To be more effective, the filler metal needs to be transferred from the electrode to the weld pool with minimum loss due to spatter.

The metal *transfer modes* during the GMAW process are described by the following categories.

1. Free-Flight Transfer
   (a) Globular
   (b) Spray Transfer
   (c) Combination of Globular and Spray

2. Short Circuiting (or Dip) Transfer

3. Pulsed Transfer

The physics of metal transfer is not yet well understood, due to the facts that the arcs are too small, the temperatures are too high, and the metal transfer is at too high a rate. Thus, many mechanisms affecting metal transfer have been suggested in [14]. To better understand the dynamics of the metal transfer in GMAW, Kovacevic [73] has studied the dynamic modeling of metal transfer process in GMAW, and presented a model for process control.

Since a stable metal transfer mode is usually associated with a good quality weld, Quintino [74] has investigated the control of metal transfer in GMAW using neural networks. This has been done through the use of sensors, control algorithms, and neural networks.

Various welding parameters, as well as the material properties of the electrodes need to be taken into consideration simultaneously for the analysis of the metal-transfer phenomena of GMAW. In order to determine the dominant factors that affect the metal transfer mode, Choi [75] has conducted a study for dimensional analysis of metal transfer.

2.4.1 Globular Transfer

Here during the welding process, the filler metal transfers across the arc in globules propelled by arc forces. As shown in Figure 2.4 for GMAW, globular transfer (characterized by a drop size bigger than the diameter of the electrode wire and at the rate of few drops per sec) takes place when the current is relatively low compared to the current levels associated with spray transfer but larger than the current levels associated with short circuit transfer. Globular transfer is also called *open-arc welding* which can take place when using a non-pulsed current source. Further works on globular transfer can be found in [76].

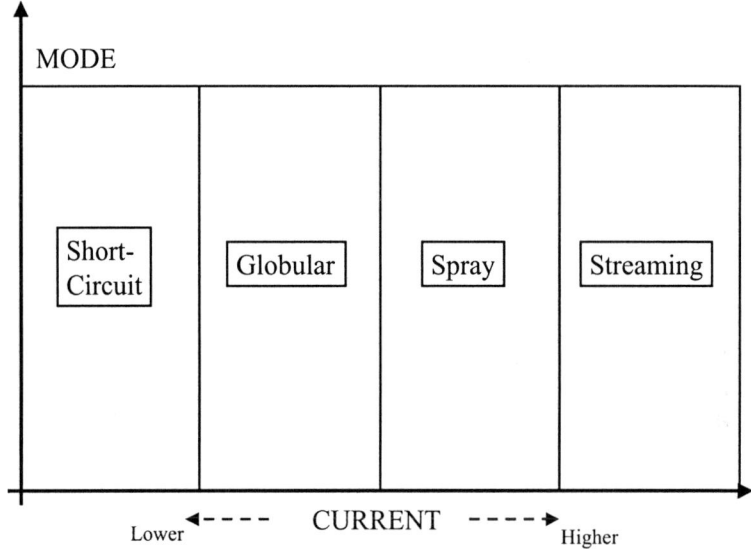

Figure 2.4: Transfer modes as a function of current.

A mathematical model to describe the globular transfer in GMAW is developed by Fan [77]. This work is both theoretical and experimental. Using the volume-of-fluid (VOF) method, the fluid-flow and heat transfer phenomena during the impingement of a droplet on a solid substrate, arc striking, the impingement of multiple droplets on the

2.4. METAL TRANSFER CHARACTERISTICS

molten pool, and, finally, the solidification after the arc extinguishes, are studied.

In [78], a dynamic two-dimensional arc model was used to investigate the effects of the various forces acting on the droplet in GMAW. The model is based on the equations of conservation of mass, energy, momentum and current, Ohm's law and Maxwell's equations.

2.4.2 Spray Transfer

Spray transfer takes place when the current is above the level of *transition current* as shown in Figure 2.4. The transition current is directly proportional to the liquid metal surface tension, but inversely proportional to the electrode diameter. An interesting characteristic of spray transfer is the phenomena of *finger* penetration. Due to its highly directed stream, spray transfer can often be used in any position. That is, the workpiece is not required to be below the torch. Also, pulsing of the welding current (see Figure 2.5) is often used for spray transfer processes. Pulsed spray transfer using low background voltage and high peak current so that the average current is less than typical non-pulsing spray transfer is addressed in [79].

2.4.3 Streaming Transfer

At higher welding currents, the drop size decreases and the electrode tip becomes tapered and a very fine stream of droplets is projected axially through the arc leading to *streaming transfer*. This transfer is seen with high-resistivity and small-diameter wires operating at welding currents above 300 A [67].

2.4.4 Short-Circuiting Transfer

Short circuiting transfer occurs during the lowest ranges of current. Metal is transferred from the electrode to the workpiece by *direct contact* between the electrode and the weld pool at the rate of 20 to 200 times per second. Short-circuiting transfer arising in the GMAW process was analyzed using the voltage and current signals in [80, 81, 82], where it was found by theory and experimentation that the factors determining reliable monitoring of the short-circuiting transfer are short-circuiting frequency, arcing/shorting period ratio and metal back distance variations.

2.4.5 Pulsed Current Transfer

In consumable electrode welding processes, metal transfer can be achieved by using power supplies that pulse the current back and forth between the globular and spray-transfer current ranges as shown in Figure 2.4. The principle is that metal transfer from the electrode is achieved in two ways depending upon the welding current: one at a current below a certain critical current producing a globular mode (less than 10 drops per second) and the other at a current above the critical current producing spray mode (a few hundred drops per second). This critical value of the current is called *transition current*. The minimum current during the globular region is called *background current*. Figure 2.5 depicts the key characteristics of the current signal in pulsed current welding.

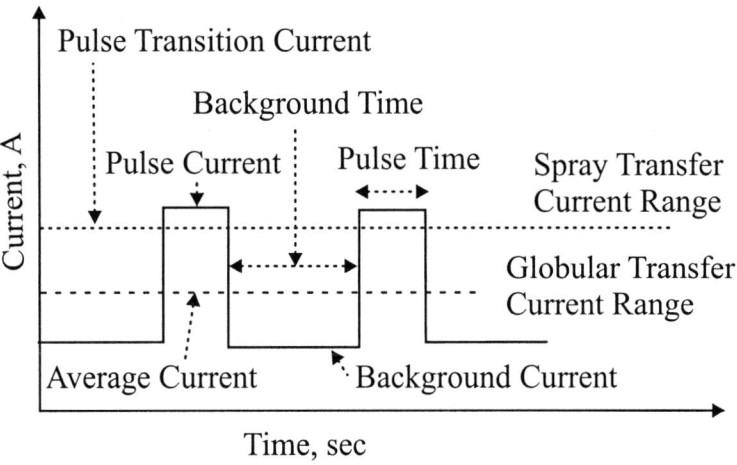

Figure 2.5: Pulsing current for metal transfer.

2.4.6 Other Works on Metal Transfer

An extensive literature survey of research results on arc physics and metal transfer up to February 1942, done in two parts (one connected with low-amperage arcs and the other related to welding arcs) is presented in [84]. One of the earliest investigations on different modes of metal transfer with particular reference to welding of aluminum can be found in [85].

A good discussion on pulsed power supplies for GMAW is given in [86, 87] and Cram [88] gives a general discussion on droplet formation. An in-depth study on pulsed GMAW with respect to the effects and interactions for some of the possible variables, including experimentation and the influence of shielding gases and wire size on bead characteristics, is presented in [89, 90].

A synchro-pulse GMAW method of pulsed power welding, where the arc length is held constant by using arc voltage as the reference for a feedback control system, can be found in [91]. The use of a power supply for investigating wire melting rate and metal transfer behavior in steady and pulsed current modes for a GMAW of aluminum is given in [92]. In [93], a study on metal transfer in pulsed GMAW with mild steel electrode considered the variations of transfer modes with various waveforms and the calculation and experimental verification of the profile of a pendant droplet at the electrode tip taking into account surface tension, gravity, and electromagnetic force.

An experimental study of pulsed-current GTAW to determine the effects of frequency on the arc column and weld pool is presented in [94]. This shows that high-frequency current pulsing increases the electromagnetic string force in the weld pool and therefore, increases weld depth independent of top width, resulting in different pool shapes. A detailed discussion on droplet transfer in GMAW is given in [95]. A study of droplet rates for various metal transfer modes in GMAW reported that at the maximum droplet rate, the droplet transfer cycle time was very consistent [96].

Different calorimetric and temperature measurements have been done [97] to determine the droplet heat content during GMAW of aluminum and the effect of droplet transfer mode and the welding current on droplet heat content and temperature were also observed. A study on power inputs to arc and cathode heating can be found in [98]. An analysis of three major types of metal transfer modes in SMAW (explosive transfer, short-circuiting transfer, and slug-guided transfer) is given in [99]. In [100], the analysis of metal transfer in GMAW is discussed. The droplet sizes are predicted using both static force balance theory and pinch instability theory as a function of welding current. Experimentation results are given.

Both theoretical and experimental results on droplet growth and detachment for globular transfer mode, the transition from globular mode to streaming mode, and streaming mode transfer are found in [101]. The development of an interesting two-dimensional time-dependent model for the prediction of droplet formation in GMAW by presenting a unified treatment of the arc welding wire (taken as the anode and the workpiece as the cathode) is given in [102]. Detailed analysis of metal transfer for GMAW control can be found in [103, 104].

2.4.7 Metal Transfer Experiments

In a series of experiments, a research team at the Idaho National Engineering and Environmental Laboratory (INEEL) investigated the various modes of metal transfer. In particular, a series of experiments was performed from globular to spray transfer at a variety of conditions using an elaborate data acquisition system that included high-speed movies [105, 106, 107, 108, 109]. It was found that at the boundary between the globular and spray transfer modes, the droplet size varies between small droplets, which melt off faster than average resulting in a smaller electrode extension and a longer arc length, and large droplets, which melt off slower than average resulting in a large electrode extension and shorter arc length. In particular, in [105], a multisensor approach was taken to determine the metal transfer mode by monitoring current and voltage of the power supply. Also, a portable workstation based on a MicroVaX II Computer and Computer Automated Measurement and Control (CAMAC) was used. In [106], experimental techniques were developed for the measurement of heat transfer from the torch to weldment and for the measurement of droplet transfer characteristics including droplet transfer time, droplet volume, and droplet velocity. Further investigations into the control of drop detachment in GMAW can be found in [110]. In this work, it was shown that complete decoupling of the metal transfer processes and base metal

2.4. METAL TRANSFER CHARACTERISTICS

heating process were possible by using independent power supplies to the system.

Employing an on-line data acquisition GMAW system using ER 100S-1 electrode, several parameters were identified to characterize different metal transfer modes [111, 112]. The analysis involved statistical methods, Fourier transforms, amplitude frequency histograms, peak searching algorithms, and smoothing techniques.

Subramaniam [113] has identified multiple transfer modes and characterized the conditions under which they occur. In the pulsed GMAW process, it was found that droplet transfer mode is affected not only by welding voltage and current, but also by the pulsing parameters. Other experimental works on droplet detachment are reported in [114, 115].

Investigation of the interaction of a molten droplet with the liquid weld pool surface has been reported in [116]. A finite element fluid dynamics model that incorporates nonlinear temperature and compositionally-dependent material properties was used. The effects of single and successive droplet transfer, droplet composition temperature, velocity, size, and transfer rate were investigated.

2.4.8 Physics of Metal Transfer

Metal transfer dynamics are the result of a balance of forces acting on the metal droplet [67]. The various forces involved are

1. Gravitational force, F_g,

2. Aerodynamic (drag) force, F_d,

3. Electromagnetic force, Fe,

4. Surface tension force, F_s,

5. Vapor jet forces, F_v,

These forces depend on particular operating conditions (weld current, arc voltage, wire diameter, shielding gases, etc.). The balance of forces on a droplet is given by

$$F_g + F_d + F_e = F_s + F_v \qquad (2.11)$$

Following is a brief description of these forces with particular with respect to free-flight metal transfer based on [67] (an alternative treatment of the various forces is found in [117]).

Gravitational Force

The gravitational force is given by

$$F_g = gm \qquad (2.12)$$

where m is the mass of the droplet and g is the vertical component of the acceleration due to gravity (9.81 m/sec^2 or 9.81cosθ, where θ is the angle between the arc axis and the vertical).

Aerodynamic (Drag) Force

The gas flow (atmosphere) around and within the arc induces aerodynamic drag (F_a) on the droplet given by

$$F_d = 0.5\pi V^2 \rho r^2 C_d \qquad (2.13)$$

where V is the gas velocity, ρ is the gas density, r is the droplet radius and C_d is the drag coefficient. This force is higher with higher droplet radius and gas velocity.

Electromagnetic Force

It is well known that a current-carrying conductor establishes a magnetic field (hence a force) around the conductor. Due to the welding current, the electromagnetic force is given by

$$F_e = \frac{\mu}{4\pi} I^2 \ln\left|\frac{r_a^2}{R}\right| \qquad (2.14)$$

where μ is the magnetic permittivity of the material, I is the welding current, r_a is the *exit* radius of the current path and R is the *entry* radius of the current path [67].

Surface Tension

Surface tension is important in metal transfer. In free-flight transfer, it is the principal force that prevents droplet detachment and in dip (short-circuit) transfer, it is the force that pulls the droplet into the weld pool. Using the static analysis of the drop retaining force in globular transfer, we get the surface tension (force) as

$$F_s = 2\pi r_w \sigma f(r_a/c) \qquad (2.15)$$

where r_w is the wire radius, σ is the surface tension, $f(r_w/c)$ is the function of wire diameter and c is the constant of capillarity. For large diameters of droplets, this force becomes

$$F_s = 2\pi r_w \sigma. \qquad (2.16)$$

Vapor Jet Forces

At higher welding currents, significant vaporization at the surface of the molten droplet can occur in the arc root area. Thermal acceleration of the vapor particles into the arc plasma results in a force called the *vapor jet force*, which opposes the droplet transfer. This vapor jet force for a flat surface at uniform temperature and composition is given by

$$F_v = \frac{m_0}{d_f} I J \qquad (2.17)$$

where m_0 is the total mass vaporized per second per ampere, I is the current, J is the current density, and d_v is the vapor density.

2.5 Weld Pool

In all arc welding processes, formation of a weld pool is a common characteristic. Upon solidification of the weld pool, the liquid metal converts into a *weld bead*. One of the most important factors determining quality of weld is the degree of penetration, which is defined as the relative pool depth, compared to the thickness of the workpiece [63]. Depending upon the penetration depth being equal to, less than, or more than the workpiece thickness, we have the full or partial penetration [118].

2.5.1 Weld Pool and Weld Bead Geometry

The weld pool geometry is the shape of the pool for a given weld pool mass and is determined by the heat conduction of the metal and the liquid motion (convection) in the weld pool [63]. The geometry of the weld pool and the weld penetration are influenced mostly by weld pool convection. In looking at the geometry modeling of the weld, Doumanidis [119] classifies the welding characteristics based on the following parameters:

1. Weld bead geometry

2. Microstructure and material

3. Residual stress and distortion

An analytical model was obtained in the solid region for the conduction temperature field in terms of pool width, penetration depth, reinforcement height, and other variables of the process. Doumanidis's design of a geometry control system is based on a linearized dynamic experimental model, obtained by an off-line process identification method in the neighborhood of the nominal conditions.

Thermal models of the welding process can be classified into three types:

1. empirical models,

2. finite element models, and

3. analytical models.

Of these, analytical models are the most promising for a control scheme. The earliest analytical model of welding may be the work by Rosenthal for point source solution of temperature in the weldment [32, 120]. These models were further refined in [37]. A distributed source conduction model is presented in [121] for prediction and control of weld geometry (weld width and depth). A real-time calibration of the model in a closed-loop control environment is also discussed.

Andersen et al. [122, 123, 124] obtained a weld pool dynamic model based on the assumption of spherical or nearly spherical geometry as in a fluid drop, different from the previous researchers who assumed a

2.5. WELD POOL

cylindrical geometry [125, 126, 127, 128]. The model is derived based on the Legendre polynomial formulation of the differential equations describing the pool dynamics. The frequency f of oscillation (for the first mode) of the weld pool is obtained as

$$f = \sqrt{\frac{T_s}{\frac{3}{8}\pi m}} \qquad (2.18)$$

where T_s is the surface tension of the molten metal in weld pool and m is the mass of the molten metal. Thus, the weld pool frequency depends on the mass and not the shape of the weld pool.

Kim [129] developed a mathematical model for control of weld beam penetration in the GMAW process. Welding process parameters included in the modeling are gas flow rate, welding speed, arc current, and welding voltage. Experimental results have shown that weld bead penetration increased as wire diameter, arc current, and welding voltage increased, whereas an increase in welding speed decreased the weld bead penetration.

A static equilibrium model for bead formation in high-speed GMAW is presented in [130]. This model is based on the Young-Laplace equation that describes the surface shape of the weld bead. Numerical simulation of a time-dependent three-dimensional GMAW pool due to a moving arc has been performed in [131]. A transient three-dimensional GMAW pool was simulated numerically. The addition of molten material was modeled by an impacting liquid metal spray on the weld pool, with evaporation and latent heat absorption for boiling being computed at the weld pool surface.

Modeling of the weld pool behaviors in GMA welding has also been developed in [132]. In this work, a new distribution mode for the arc heat flux on the deformed weld-pool surface has been established according to the physical phenomenon of the arc current conducting path, which is modified by the deformation of weld-pool surface.

2.5.2 Other Works on Weld Pool

The development of a model for the dynamics of full penetration that relates the heat input to the width of the back bead for GTAW process is given in [133, 134]. A study on improving the weld quality (structure and mechanical properties) of aluminum welds by low frequency arc oscillation during GTAW is presented in [135]. A new model of a compound vortex as a possible mechanism to explain the deep surface depression arising at currents over 300 A using calculus of variations is proposed in [136]. A new system called MELODY has been developed by the Machwood Engineering Laboratories, for weld pool frequency monitoring [126].

Tamhardt [137] has developed a model for pool motion applicable to both stationary and non stationary weld cases for a GTAW process, where the models relate the pool geometry parameters to frequency characteristics of the pool motion. Identification of the weld pool impedance is discussed in [138]. In [139], two models have been developed for weld pool. The first one is a lumped parameter model and the second one is a distributed parameter model. Weld pool geometry is predicted based on the natural frequency of the weld pool, where the weld pool oscillations are measured through signal processing of arc voltage and current signals for a stationary GTAW process.

The derivation of a lumped dynamic model of the weld bead geometrical characteristics (width, penetration and reinforcement height) for partial penetration GTAW processes is presented in [140]. A hybrid model for the weld geometry consisting of three parts: a torch model, the weld pool, and the solid region was developed by Doumanidis [141].

In [142] an experimental study on weld pool oscillations or weld pool geometry using arc voltage and arc light emission for a GTAW process was reported. Investigations on three-dimensional heat transfer and fluid flow in GMAW for analyzing the effect of contact tube-to-workpiece distance on the weld pool geometry by considering the driving forces for weld pool convection, electromagnetic force, the buoyancy force and the surface tension force on the weld pool surface can be found in [143]. In this work, experimental results have shown that the variation of the weld bead geometry is due to the change of the contact tube-to-workpiece distance. Other related works are reported by the same authors [144, 145].

Further, a three-dimensional weld pool model has been developed to study the fluid flow and heat transfer during GMAW [146]. Both droplet heat content and impact force were considered in analyzing the effect of droplets on the formation of weld pool. The fluid flow in the

2.6. PROCESS VOLTAGES

weld was induced by the presence of surface tension, electromagnetic and buoyancy force. The surface deformation of the weld pool was calculated by considering arc pressure and droplet impact force.

2.6 Process Voltages

Another way of accounting for all the voltage drops in the welding circuit is given in [147]. Let the process voltage E_p be denoted as the voltage that appears across the contact tube and workpiece (or the source voltage minus the drop across the internal resistance of the source). Thus, the process voltage E_p is given by

$$E_p = E_{cath} + E_{col} + E_{anod} + E_{so} + E_{cont} \quad (2.19)$$

where the various voltages are as shown in Figure 2.6 [147].

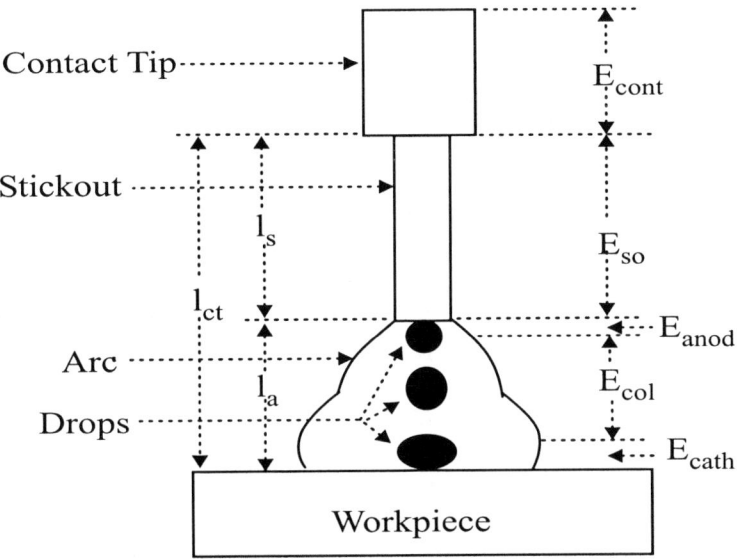

Figure 2.6: Process voltage in the GMAW process.

2.6.1 Cathode and Anode Voltages

The cathode ($Ecath$) and anode (E_{anode}) voltages result from the arc (or plasma) dynamics and are often combined [117].

2.6.2 Arc Column Voltage

As it is often difficult to separate the arc column voltage (E_{col}) from the anode and cathode voltage, it is often combined with them and the three are collectively termed the "arc voltage". Thus the arc voltage may be written as

$$E_a = E_{anode} + E_{cath} + E_{col} = E_{ac} + E_{col} \qquad (2.20)$$

where E_{ac} is the combination of anode and cathode voltages. If the welding current I is nearly constant, then the arc column E_{col} is observed to be directly proportional to the arc length l_a. Then

$$E_a = E_{ac} + \gamma l_a \qquad (2.21)$$

where γ is the gradient of the arc voltage divided by the arc current, which is mainly a function of the composition of the shielding gases. Alternatively, arc voltage E_a has been determined to be a more complicated function of current [117]:

$$E_a = \gamma_0 + \gamma_1 I + \frac{\gamma_2}{I} \qquad (2.22)$$

where γ_0, γ_1, and γ_2 are constants. On the other hand, taking the arc column as a conductor of current, we can write

$$E_a = R_a I = \rho_a \frac{l_a}{A_a} I \qquad (2.23)$$

where R_a is the effective resistance of the arc, l_a is the arc length and A_a is the effective cross-sectional area of the arc.

2.6.3 Stick-Out Voltage

Stick-out voltage (E_{stick}) is also called the electrode extension voltage, which is given from the fundamental relation

$$E_{so} = R_{so} I = \rho_s \frac{l_{so}}{A_{so}} I \qquad (2.24)$$

where R_{so} is the resistance of the stick-out, K is the welding current, ρ_s is the resistivity of the stick-out material, l_{so} is the stick-out length, and A_{so} is the area of cross-section of the stick-out.

2.6.4 Contact Tip Voltage

The contact tip voltage (E_{con}) is the voltage between the electrode and the contact tip and depends on the condition of the contact tip.

2.7 Heat and Mass Transfer

2.7.1 Model for Heat and Mass Transfer

This material is from the work of Smartt and Einerson [148], in which a steady-state model of the GMAW process is obtained for electrode melting and heat and mass transfer from the electrode to the workpiece.

The electric power consumed by the process is approximately equal to the sum of that consumed by the resistive heating of the electrode and that consumed by the arc, or

$$P = IE = IV_e + IV_{arc} \tag{2.25}$$

where I is the current from the electric source, E is the secondary circuit voltage drop, V_e is the voltage drop across the electrode, and V_{arc} is the voltage drop across the arc. Next, the heat input H to the base metal per unit length of weld is given by

$$H = \frac{EI\eta}{R} = \frac{I(V_e + V_{arc})\eta}{R} \tag{2.26}$$

where η is the heat transfer efficiency from the process to the base metal, and R is the weld speed. The weld reinforcement G, defined as the transverse cross-sectional area of the deposited metal, is given as

$$G = \frac{AS}{R} = \frac{M_R}{R} \tag{2.27}$$

where A is the cross-sectional area of the electrode, S is the electrode-feed speed, and M_R is the melting rate. These above relations will be used later on in obtaining the necessary control actions for heat and mass transfer.

Since the mass G and heat H transferred to the workpiece determine the quality of welding, it is necessary to investigate the relation between $G - H$ and the welding parameters. In other words, determine what the welding parameters are for given desired $G - H$ values. Details of this topic are given in Chapter 5 and by Ozcelik, et al. [149].

2.8 Process Variables

According to Dornfeld, et al. [150], GMAW process variables can be categorized into three types:

1. Variables that can be varied on-line during the process.

2. Variables that are set prior to the beginning of the process.

3. Variables that cannot be modified.

For the GMAW process, the key process variables of the first category are the power supply voltage and polarity, the wire-feed rate and the resulting current, and the torch speed; the key variables that are set before the start of the process are the shielding gas flow and composition, the torch angle, the contact tube-to-workpiece distance, and the electrode material and size; the key variables belonging to the third category are the workpiece (plate) thickness, the joint geometry, and the physical properties of the base metal.

The weld penetration, bead geometry, and quality of weld are affected by

1. Welding current (electrode-feed rate).

2. Polarity of power supply.

3. Arc voltage (arc length).

4. Weld torch travel speed.

5. Electrode extension (stick-out).

6. Electrode diameter.

7. Electrode orientation.

2.8. PROCESS VARIABLES

8. Shielding gas composition and flow rate.

In the remainder of this section we will briefly discuss these variables.

2.8.1 Welding Current

Keeping all other variables constant, welding current varies directly with electrode wire feed speed (rate) (WFS) or melting rate as

$$WFS = aI + bl_{so}I^2 \qquad (2.28)$$

where WFS is the electrode wire feed speed; a is a proportionality constant for anode or cathode heating, whose magnitude is dependent upon polarity, composition, and other factors; b is another proportionality constant for electrical resistance heating; l_{so} is the electrode extension or stick-out, and I is the welding current).

Note that as the electrode wire diameter is increased (while maintaining the same electrode wire-feed speed), higher welding current is required. Also, higher welding current (with all the other variables being kept constant) results in

1. Higher deposition rates.

2. Increased depth and width of weld penetration.

3. Increased size of the weld bead.

The influence of arc current on the mechanical properties of weld joints are experimentally reported in [151].

2.8.2 Polarity

The term *polarity* is used with the type of electrical connection of the welding gun with the electrical power supply. With gun power terminal connected to the positive terminal of the power supply, it is called *direct current electrode positive (DCEP)* or *straight polarity*, which provides desirable features of stable arc, smooth metal transfer, low spatter, good weld-bead, and good penetration. *Direct current electrode negative (DCEN)* connection is rarely used since the drop size is large and the arc forces propel the drops away from the workpiece leading to the arc becoming unstable.

2.8.3 Arc Voltage (Arc Length)

Arc length and arc voltage are often used interchangeably although they are different but related. Arc length is an important variable in GMAW process. It is important to note that arc length is an independent variable and arc voltage is dependent on arc length as well as on many other variables. Arc voltage often includes the voltage drop across the electrode extension (stick-out). With all other variables kept unchanged, the arc voltage is directly related to arc length.

2.8.4 Travel Speed

Travel speed is the rate at which the welding gun is moved in the direction of welding or along the weld joint. With low travel speeds, the filler metal deposition is high and at very low speeds the welding arc impinges on the molten pool rather than the workpiece. With increased travel speeds, the thermal energy per unit length of weld transmitted first increases and then decreases. At very high speeds, there is insufficient deposition of filler metal. Hence, with all other things being the same, the weld penetration is maximum at some moderate speeds of the weld torch.

2.8.5 Electrode Extension (Stick-Out)

The stick-out or electrode extension is the distance between the tip of the contact tube and the end of the electrode. An increase of electrode extension obviously increases the electrical resistance of the circuit, which in turn increases the heating and hence the temperature, leading to increased melting rate. But on the whole, this increased resistance absorbs higher voltage across it and as a result, the power supply decreases the welding current. This decreased current reduces the melting rate resulting in shorter arc length. Thus the stick-out is said to be "stabilized." The stick-out is usually in the order of 1/4 to 1/2 in. Calculations of the resistance of the wire extension using Joule heating principle are given by [152].

2.8. PROCESS VARIABLES

2.8.6 Electrode Orientation

The orientation of the welding electrode with respect to the weld joint affects the weld bead shape and penetration to a greater extent than the arc voltage or travel speed.

2.8.7 Electrode Size

A large electrode size (diameter) requires higher current for melting, other variables being the same. Higher currents also produce higher melting rates, leading to higher deposition rates.

2.8.8 Shielding Gases

The primary function of shielding gases is to protect (shield) the molten weld metal from contact with the atmosphere, thereby avoiding the formation of oxides of the metal. Further, the shielding gases and their flow rates have a great effect on arc characteristics, metal transfer mode, penetration and weld bead profile, weld speed, etc.

The main gases used in GMAW process are inert gases (argon and helium) and small quantities of oxygen or carbon dioxide. Argon, being heavier (1.4 times) than air, is most effective in shielding the arc and blanketing the weld pool area. Helium, being lighter (0.14 times) than air requires flow rates to be 2 to 3 times greater than the flow rates needed for Argon, but has a higher thermal conductivity than argon thereby producing a uniformly distributed arc plasma. Usually a mixture of Argon and Helium (50 to 75%) is used. Also, addition of 1 to 5% of Oxygen or 3 to 5% of CO_2 with Argon causes improvement in arc stability and weld appearance.

The importance of shielding gases in GMAW has been very well articulated in [153] by first noting that the shielding gases around the molten metal weld pool keep out the harmful atmospheric gases, such as oxygen, nitrogen, hydrogen, and moisture, which otherwise produce oxidation and other weld defects such as porosity, pinholes, and weld brittleness. Also, refer to [154] for a study on the effect of shielding gases in GMAW. This study concludes that by using shielding gases with lower content of oxidizing components (O_2 and/or CO_2), it is possible to influence the toughness and strength of the weld. See [155] for a study on the effect of shielding gas composition on the formation of desirable microstructure of welded metals.

2.8.9 Classification of Process Parameters

An alternate approach to classifying welding parameters or variables from that of [150] as presented above is to consider variables as either direct weld parameters (DWP) or indirect weld parameters (IWP) [156, 157, 158]. DWP are those relating to the weld reinforcement, fusion zone geometry, mechanical properties of the completed weld, weld microstructure, and discontinuities. IWP are those input variables that collectively control the DWP, such as welding equipment setpoint variables, such as voltage, current, torch travel speed, wire-feed speed, travel angle, electrode extension (stick-out), focused spot zone, and beam power (see Figure 2.7). The objective of a welder is to determine

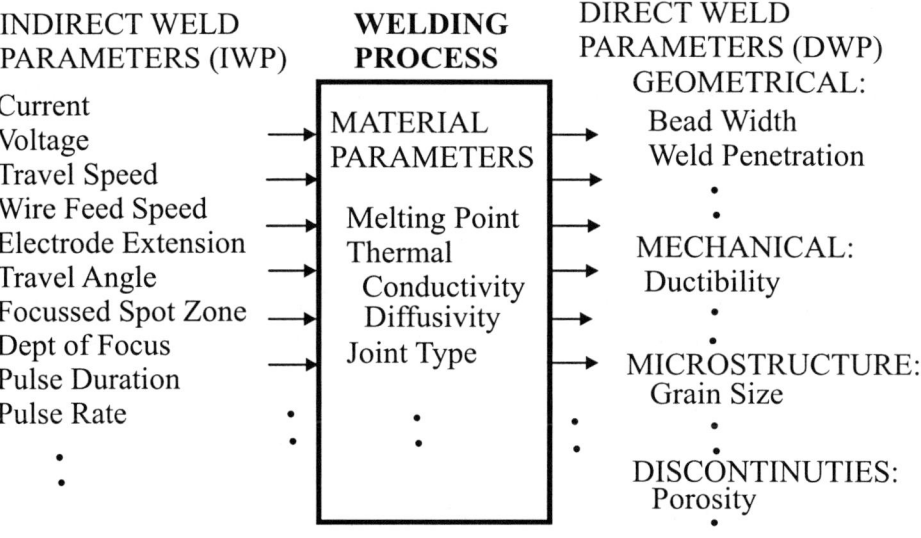

Figure 2.7: Input and output variables of the welding process.

a set of IWP that will produce the *desired* DWP. Figure 2.7 makes it clear that the welding process is a multi-input, multi-output (MIMO), multivariable system. Also, note that the relation between the input

2.9. INEEL/ISU MODEL

and output variables is dynamic, highly nonlinear, and strongly coupled. A schematic of the effect of some IWP on some DWP is shown in Figure 2.8 [159], where (+) indicates an increase is followed by an increase and (-) indicates an increase is followed by a decrease.

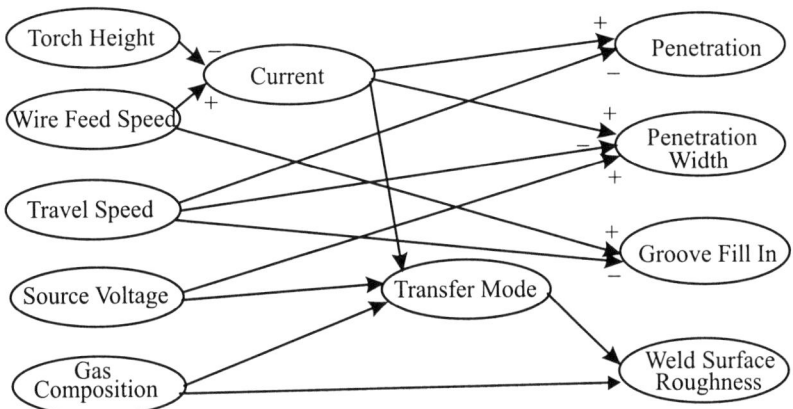

Figure 2.8: Relationship between welding parameters of the GMAW process.

2.9 INEEL/ISU Model

In this section, we briefly describe the GMAW system and its modeling as developed by a collaborative team of researchers from Idaho National Engineering and Environmental Laboratory (INEEL) and Idaho State University (INEEL/ISU) over several years [160, 161].

A schematic representing the GMAW system along with the electric power supply is shown in Figure 2.9 [162]. The power supply consists of a constant voltage source connected to the electrode and the workpiece. Wire speed S, travel speed R of the torch, open-circuit voltage V_{oc}, and contact tip to workpiece distance CT are adjusted to get the desired weld. Here, x is the distance of the center of the mass of the droplet above the workpiece.

Modeling of the GMAW process dynamics produces a fifth-order nonlinear differential equation. The model that we are using is a fourth-generation derivative that originated at INEEL and has been subsequently developed by INEEL researchers as well as ISU researchers [160]. We do not describe these results in detail but refer the interested reader to the following earlier works in GMAW modeling [109, 117, 163]. We first list the various variables used in obtaining the model equations below. We then separately give the equations describing the dynamics of the droplet and the forces acting on the drop. Next, we give a concise state-space representation of the resulting equations. We then describe the reset condition that governs droplet detachment.

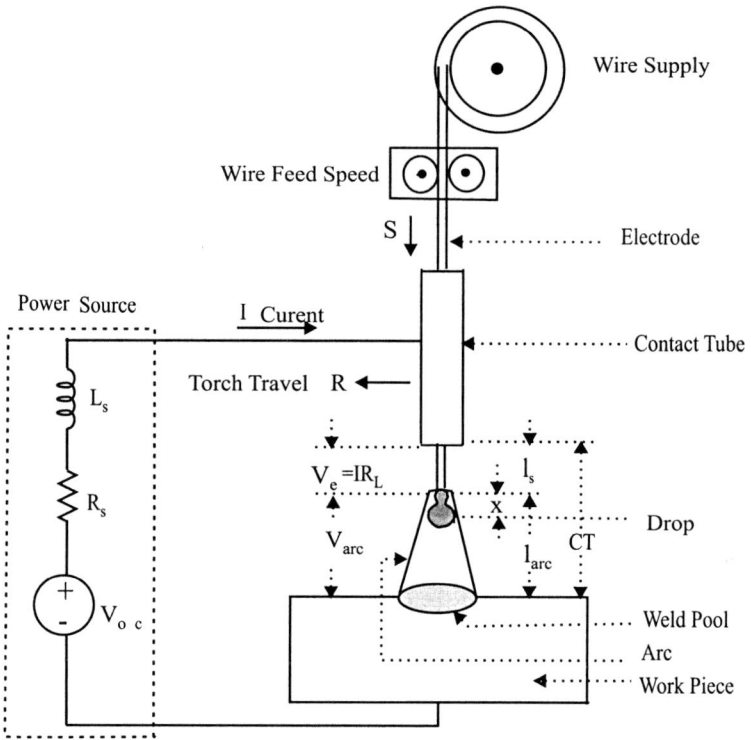

Figure 2.9: Schematic diagram of the GMAW process.

2.9.1 Nomenclature

a : Lorentz force constant = 0.641
b : Lorentz force constant = 1.15

2.9. INEEL/ISU MODEL

B : damping coefficient - $1e-5\frac{kg}{s}$
c : Lorentz force constant $= 0.196$
C_1 : melting rate constant - $2.8855e-10$
C_2 : melting rate constant - $5.22e-10$
C_d : aerodynamic drag coefficient - 0.44
CT : contact tip-to-workpiece distance - $1e-3 \leq CT \leq 0.025m$
E_a : arc length factor - $1500\frac{V}{m}$
F_d : force due to aerodynamic drag - N
F_{em} : force due to electromagnetic induction - N
F_g : force due to gravity - N
F_m : force due to momentum - N
F_s : surface tension force - N
I : current - $25 \leq I \leq 565A$
K : spring constant - $2.5\frac{N}{m}$
l_{arc}: arc length - m
l_s : stick-out - $1e-3 \leq l_s \leq CT m$
L_s : source inductance - $0.14e-3H$
m_d : droplet mass - kg
M_R : melting rate - $\frac{m^3}{s}$
r_d : droplet radius - m
r_w : electrode radius - $0.0004445m$
R : relative speed of weldment to torch - m
R_a : arc resistance - 0.022Ω
R_L : electrode resistance - Ω
R_s : source resistance - 0.004Ω
S : wire feed speed - $0.021 \leq S \leq 0.33\frac{m}{s}$
U_b : relative fluid to drop velocity - $10\frac{m}{s}$
V_{arc} : arc voltage - V
V_0 : arc voltage constant - $15.7V$
V_{oc} : open circuit voltage - $29V$
x : droplet displacement - m
$\dot{x}$: droplet velocity - $\frac{m}{s}$
$\ddot{x}$: droplet acceleration - $\frac{m}{s^2}$
ρ : resistivity of the electrode - $0.2821\frac{\Omega}{m}$
ρ_p : plasma density - $1.6\frac{kg}{m^3}$
ρ_w : electrode density - $7860\frac{kg}{m^3}$
μ_0 : permeability of free space - $1.25664(10^{-6})\frac{kg\,m}{A^2 s^2}$

γ : surface tension of liquid steel - $2\frac{N}{m^2}$

2.9.2 Forces Affecting Droplet Dynamics

We consider four forces to be acting on the droplet [117]: gravitational, electrodynamic, aerodynamic, and momentum.

Force due to Gravity

$$F_g = 9.81 m_d \tag{2.29}$$

Force due to Electromagnetic Induction (LORENTZ)

$$F_{em} = \frac{\mu_0 I^2}{4\pi} \left[\frac{a}{1 + exp\left[\left(b - \frac{r_d}{r_w}\right) \div c\right]} \right] \tag{2.30}$$

where

$$r_d = \left(\frac{3 m_d}{4\pi \rho_w}\right)^{\frac{1}{3}} \tag{2.31}$$

Force due to Aerodynamic Drag

$$F_d = \frac{C_d [r_d^2 - r_w^2] \pi \rho_p U_b^2}{2} \tag{2.32}$$

Force due to Momentum

$$F_m = M_R \rho_w S \tag{2.33}$$

where

$$M_R = C_2 I^2 \rho l_s + C_1 I \tag{2.34}$$

Surface Tension Force

The surface tension of the droplet will also be considered. It is given as

$$F_s = 2\pi \gamma r_w \tag{2.35}$$

An experimental investigation of the forces (gravity force, drag force due to flowing gas, electromagnetic force, and the retaining force by the surface tension) acting on a droplet, by measuring the drop mass as function of the gas flow rate and the electric current, are presented in [164].

2.9. INEEL/ISU MODEL

2.9.3 Droplet Dynamics

Here, the droplet is characterized as a typical spring-mass-damper mechanical system, analogous to water droplets dripping from a leaky facet [165]

$$F_{tot} = m_d \ddot{x} + B\dot{x} + Kx \qquad (2.36)$$

Rewriting the above, we have

$$\ddot{x} = \frac{F_{tot} - B\dot{x} - Kx}{m_d}$$
$$F_{tot} = F_{em} + F_d + F_m + F_g \qquad (2.37)$$

Current: The current is obtained from a simple electrical circuit principle as [162]

$$\dot{I} = \frac{V_{oc} - R_L I - V_{arc} - R_s I}{L_s}$$
$$V_{arc} = V_o + R_a I + E_a(CT - l_s) \qquad (2.38)$$
$$R_L = \rho \left[l_s + \frac{1}{2}(r_d + x) \right]$$

Stick-Out: Using the melting rate [68], we have stick-out as

$$\dot{l}_s = S - \frac{M_R}{\pi r_w^2} \qquad (2.39)$$

Mass: The mass is given as

$$\dot{m}_d = \rho_w M_r \qquad (2.40)$$

Also, see [166, 167, 168] for developing an alternative formulation for contact-tip-to workpiece distance as a function of welding current, voltage, and wire feed speed.

Detachment Criteria

- Due to force imbalance

$$F_{tot} > F_s \qquad (2.41)$$

- Due to shape instability

$$r_d > \frac{\pi(r_d + r_w)}{1.25 \left(\frac{x+r_d}{r_d}\right)\left(1 + \frac{\mu_0 I^2}{2\pi^2 \gamma(r_d+r_w)}\right)^{\frac{1}{2}}} \qquad (2.42)$$

Drop Volume

$$\text{detach volume} = \frac{m_d}{2\rho_w}\left(\frac{1}{1+\exp(-100\dot{x})}+1\right) \quad (2.43)$$

In [40], for the GMAW process, the authors conducted experimental investigations on the measurement of the velocity of drops detached from the electrode and the diameter of the drop neck during detachment using high-speed photography. It was found that the pendent drop contracts at the start of the pulse and becomes unstable when the diameter of the droplet attains some critical value. Further it was found that the drop velocity at detachment is controlled by Lorentz force [117].

A derivation of a dynamic growth and detachment droplet model for a GMAW process by using a second order spring-mass model is presented in [109] with experimental results. In [169], it was found that character of the evolution of a pendent drop at the end of a nozzle, depends strongly on the growth rate of the drop and the radius of the nozzle and that the drop becomes unstable.

2.9.4 Model Equations

We now combine the equations presented in the previous section so as to develop a state-space model. Define the *state* variables as

$x_1 = x$: droplet displacement - m
$x_2 = \dot{x}$: droplet velocity - m/sec
$x_3 = m_d$: droplet mass - kg
$x_4 = l_s$: stick-out - m
$x_5 = I$: current - A

Then the nonlinear state equations can be written as

$$\begin{aligned}
\dot{x}_1 &= x_2 \\
\dot{x}_2 &= \frac{-Kx_1 - Bx_2 + F_{tot}}{x_3} \\
\dot{x}_3 &= M_R \rho_w \\
\dot{x}_4 &= u_1 - \frac{M_R}{\pi r_w^2} \\
\dot{x}_5 &= \frac{u_2 - (R_a + R_s + R_L)x_5 - V_0 - E_a(CT - x_4)}{L_s}
\end{aligned} \quad (2.44)$$

2.9. INEEL/ISU MODEL

Melting rate M_R and the electrode resistance R_L are given by

$$M_R = C_2\rho x_4 x_5^2 + C_1 x_5$$
$$R_L = \rho\left[x_4 + \frac{1}{2}\left(\left(\frac{3x_3}{4\pi\rho_w}\right)^{\frac{1}{3}} + x_1\right)\right] \quad (2.45)$$

The output equations are

$$y_1 = V_0 + R_a x_5 + E_a(CT - x_4)$$
$$y_2 = x_5 \quad (2.46)$$

where the output variables are

$y_1 = V_{arc} = V_0 + R_a I + E_a(CT - x_4)$: arc voltage (V)
$y_2 = I$: current (A)

and the control variables are

$u_1 = S$: wire-feed speed - m/sec;
$u_2 = V_{oc}$: open-circuit voltage - volts;
$u_3 = CT$: contact tip-to-workpiece distance - m;

Reset Conditions

States of the system must be reset after each detachment of a drop takes place. Therefore, whenever

$$F_{tot} > F_s \quad (2.47)$$

or

$$r_d > \frac{\pi(r_d + r_w)}{1.25\left(\frac{x+r_d}{r_d}\right)\left(1 + \frac{\mu_0 I^2}{2\pi^2\gamma(r_d+r_w)}\right)^{\frac{1}{2}}} \quad (2.48)$$

where

$$r_d = \left(\frac{3x_5}{4\pi\rho_w}\right)^{\frac{1}{3}} \quad (2.49)$$

the volume of the droplet V_d that detaches (based on a function of drop velocity) is taken to be

$$V_d = \frac{m_d}{2\rho_w}\left(\frac{1}{1+\exp(-100\dot{x})} + 1\right) \quad (2.50)$$

After each detachment, the states are updated as follows

$$\begin{aligned} x_1 &= r_d \\ x_2 &= 0 \\ x_3 &= \frac{x_3}{2}\left(1 - \frac{1}{1+\exp(-100x_2)}\right) \\ x_4 &= x_4 \\ x_5 &= x_5 + \frac{u_2 - V_0 - E_a(CT - x_4)}{R_a + R_s + R_L} \\ &\quad - \frac{u_2 - V_0 - E_a(CT - x_4)}{R_a + R_s + R_{Lold}} \end{aligned} \quad (2.51)$$

where R_{Lold} is the value of R_L before detachment. Otherwise,

$$\begin{aligned} x_1 &= x_1 \\ x_2 &= x_2 \\ x_3 &= x_3 \\ x_4 &= x_4 \\ x_5 &= x_5 \end{aligned} \quad (2.52)$$

Combining all these equations, the model of the GMAW process can be written in the following general form:

$$\begin{aligned} \dot{\mathbf{x}} &= \mathbf{f}(\mathbf{x}) + \mathbf{g}(\mathbf{x})\mathbf{u} \\ \mathbf{y} &= \mathbf{h}(\mathbf{x}) + \mathbf{i}(\mathbf{u}) \\ \mathbf{x} &= \mathbf{k}(\mathbf{x}) \quad \text{if} \quad L(\mathbf{x},\mathbf{u}) > 0 \end{aligned} \quad (2.53)$$

In [163, 101], ISU and INEEL researchers developed the above model for the GMAW process based on some physical phenomena of the process. Further, the model was calibrated with experimental results obtained at the INEEL facility. The experimental hardware from Advanced Manufacturing Engineering Technologies (AMET) was used to have a computer-controlled welding of a rotating flat plate. Several other advanced techniques and equipment were used in the experimental facility. Other related works can be found in [109, 108].

2.9.5 Model Simplification and Linearization

Let us note that the GMAW model dynamics as described by the fifth-order differential equations (2.45) are highly *nonlinear*, which makes it

2.9. INEEL/ISU MODEL

difficult for many of the modern control strategies to be used successfully. To help overcome this, we present a simplified model based on some approximations. First, let us rewrite the current I and stick-out l_s relations from (2.45) as

$$\dot{x}_4 = u_1 - \frac{M_R}{\pi r_w^2}$$
$$\dot{x}_5 = \frac{u_2 - (R_a + R_s + R_L)x_5 - V_0 - E_a(CT - x_4)}{L_s} \quad (2.54)$$

where

$$R_L = \rho(x_4 + 0.5(r_d + x_1)) \quad (2.55)$$

and $CT = u_3$ is assumed to be constant, which facilitates two-state, two-input system instead of a two-state, three-input system. Now, we make the following valid approximation that the stick-out distance ($l_s = x_4$) is much larger than the sum of the droplet radius r_d and the drop distance x_1. That is

$$x_4 \gg 0.5(r_d + x_1) \quad (2.56)$$

Note that pictures obtained using a high-speed camera show the validity of this approximation (see [170], for example). Using the above simplification in the current relation (2.54), we have

$$\dot{x}_4 = u_1 - \frac{M_R}{\pi r_w^2}$$
$$\dot{x}_5 = \frac{u_2 - (R_a + R_s + \rho x_4)x_5 - V_0 - E_a(CT - x_4)}{L_s} \quad (2.57)$$

which can be written as

$$\dot{\mathbf{x}}(t) = \mathbf{A}\mathbf{x}(t) + \mathbf{f}(\mathbf{x}) + \mathbf{B}\mathbf{u}(t) \quad (2.58)$$

Linearization of the above equation about steady-state is used to obtain adaptive control of the GMAW process [149]. This is described in Chapter 5. Alternate derivations of the model using the weld current I and stick-out l_s are given in [171].

2.10 Empirical and Statistical Models

In practice, the most common technique to control the welding process variables is to obtain a model based on some empirical data. For example, one can conduct a number of experiments to determine the effect of a certain input variable, say on the weld bead geometry, and then mathematical expressions can be found to fit the experimental data [172, 173, 174, 175, 176, 177]. It so happens that due to the complexity of the welding process, one has to perform a number of experiments. However, using statistical analysis, it may be possible to reduce the number of experiments [178].

In [179], an efficient method of solving the problem of rational selection of welding conditions using mathematical methods of multicriteria optimization based on statistical models of the welding process is given. A formal description of the dependence of joint properties on welding parameters treated as regression models makes it possible to derive an automatic control system for the process.

Semi-empirical models of the fume formation in GMAW are discussed in [180]. The work examines the fundamentals of welding-fume formation. A physical chemistry model of the metal vapor mechanism for fume formation has also been developed for nonshort-circuiting transfer GMAW. The model includes the important contribution made by direct condensation of metal vapor onto the weld pool and the workpiece in removing a substantial fraction of the fume.

Kang [181] developed statistical models for estimating the amount of spatter quantitatively using the wave forms in the short circuit transfer mode of GMAW. In this study, the spatter was gathered under several welding conditions and, at the same time, the waveforms were measured. The factors representing the characteristics of the waveforms were calculated from the measured waveforms. Four different linear and nonlinear regression models were proposed to estimate the amount of spatter based on a multiple regression analysis between each model and the amount of spatter.

2.11 Modeling by System Identification and Estimation

Due to the complexity of the GMAW process, Nishar, et al. [182] used a standard system identification (ID) and estimation (EST) for model rather than deriving the model from the first principles of physical laws of the process. Data was obtained for a set of experiments, and the averaged set of data was used to obtain a preliminary model by using recursive least squares estimation. This model was further refined by using the approximate maximum and minimum likelihood algorithms. Although the procedure resulted in system models of order 1 to 10, it was decided to use a second-order system that gave one of the lowest mean square error.

In [183] a linear system identification scheme using the input-output data in the absence of a mathematical model was proposed for the GMAW process where the input is a welding current and output is depth, heat-affected zone width, and pool width.

Distributed Parameter Model

In order to have combined control of structure, properties and stress conditions of the weld material and to estimate the internal thermal field, a distributed-parameter model was derived in [119, 184].

2.12 Intelligent Modeling

Here we present modeling techniques using artificial neural networks (ANN), fuzzy logic, and expert and knowledge-based systems. Modeling geometrical parameters of a GTAW process using neural networks is reported in [185]. A modified back-propagation network to include state information of the dynamic system was given in [186]. These results showed good agreement between the ANN model and the experimental model.

Andersen [187] addressed the application of an ANN to model the map between the indirect welding parameters such as arc current, travel speed, arc length and plate thickness and the direct weld parameters (DWP) such as weld-bead width and weld penetration of GTAW. In particular, the ANN with back-propagation contained 10 nodes in a single hidden layer.

2.12.1 Other Works on Intelligent Modeling

An excellent discussion on modeling, sensing, and control of welding processes is given by [157]. Kuvin [188] presented a program to revamp welding of US Army tanks by General Dynamics using GMAW at the rate of 100,000 inches of weld length, 1500 pounds of metal deposit and 200 hours of arc time per tank, 720 tanks/year.

Artificial intelligence (AI) methods (in particular, neural networks) for process modeling are discussed in [189]. Cook [190] used two-hidden layer ANN for modeling a GTAW process, with one ANN being used for mapping between desired output features and required equipment variables and another ANN being used for estimating output features that are not directly sensed. Another application of an ANN to welding process parameter modeling is given by [191].

In [192, 193], a systematic on-line method for obtaining the proper voltage/current combination to provide a stable arc condition in the short-circuiting metal transfer mode of a GMAW process is proposed. This method uses fuzzy rule-based linguistic rules to represent the complexity and nonlinearity of the arc behavior, the two fuzzy variables used being Mita's arc stability index and its derivative with respect to voltage.

The design of a neural network estimator (NNS) for deriving the weld pool size from surface temperatures measured at various points on the surface of the weldment in a GMAW process is given by [194]. The input to the NNS is the multi-point temperatures at the top-side of weldments which are measured by an infrared temperature sensing system, and the outputs of the NNS are weld geometrical parameters such as top bead width and penetration plus half-back bead width. The NNS is shown in Figure 2.10

Modeling gas metal arc weld geometry using ANN technology is discussed in [195]. A back propagation network system for predicting gas metal arc (GMA) bead-on-plate weld geometry from current, voltage, and wire travel speed is reported in this study. Moreover, work-piece thickness is a variable that is taken into consideration because its effect on weld shape is at this stage unknown in practice. The database

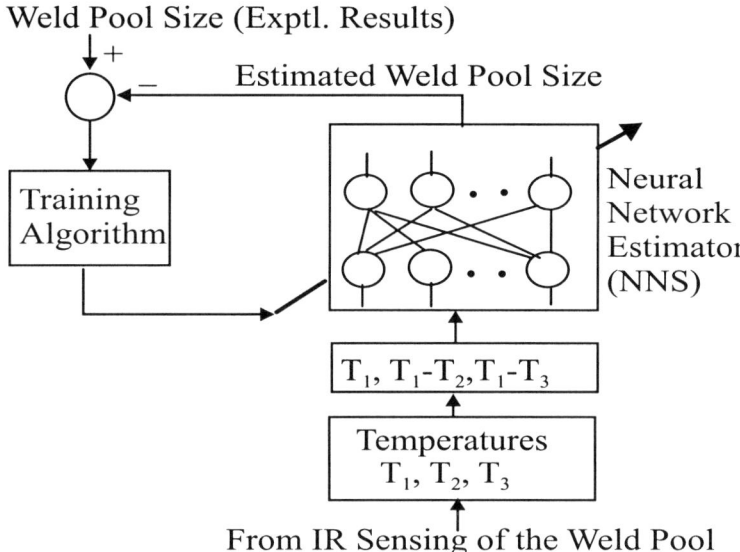

Figure 2.10: Neural network estimator for weld pool geometry.

consists of some ninety six welds.

Static modeling of the GMAW process using ANN is given by Di [196]. In this study, due to the complexity of the welding process, neural network-based approaches have been considered as an effective way to represent the required model.

2.13 Other Issues on Modeling

2.13.1 Dawn of GMAW

There is a very interesting article [197] on the recollections of Glen Gibson, one of the three inventors of U.S. Patent 2,504,688 issued April 18, 1950, which is considered by most experts as the basic GMAW patent and the beginning of the modern welding system using the continuous consumable electrode welding process.

2.13.2 Cost of GMAW

The cost of applying weld metal using the GMAW process consists of four major items: labor and overhead, electrode, shielding gas, and electrical power. For details of these four items refer to [198]. Return on investment for an automated arc welding system is discussed in [199].

Edgards [200] gives suggestions regarding controlling the metal volume and hence improving productivity and profit in a particular welding process, such as holding fillet welds to specified size, depositing equal-leg fillets, minimizing weld face reinforcements, minimizing root opening and proportioning groove angles, and finally saying "do not over specify and do not overweld." A discussion on quality and economy, weld cost analysis, and analysis of individual costs in active GMAW is given by [201].

2.13.3 Other Works on Modeling

The influence of welding parameters on droplet temperature during pulsed arc welding is discussed in [202]. A systematic technique for the analysis of numerical data and its application to the selection of four different arc welding processes is presented in [203]. McGlone has proposed an approach in [172] for procedure selection in arc welding.

An interesting article on "how difficult is it to learn gas metal arc welding?" [204] says that learning GMAW can take over 30 hours of training. A study on fume generation and melting rates of shielded metal arc electrodes is given in [205]. In this study, it was found that the fume generation rates (g fume/kg electrode melted) increased almost linearly with voltage and power and decreased almost linearly with current.

A GMAW process was developed at the Westinghouse R&D Facility, Pittsburgh, PA. It added a second power source that preheats the filler wire before it emerges from the welding torch. This made it possible to break the fixed relationship that for any combination of filler wire and power source, the welding current, electrode extension and deposition rate cannot be individually controlled, thus providing independent and balanced control of the variables so as to make the new process extremely versatile and flexible for welding operations [206, 207].

2.13. OTHER ISSUES ON MODELING

The quantitative relationship of the surface weld bead with stability of arc and uniformity of droplet detachment in pulsed GMAW was studied in [208]. It was shown that high arc stability and uniform drop detachment provide a smooth weld bead surface. Arc control experiments in GMAW, plasma arc welding and shielded metal arc welding in China are presented in [209].

Investigations into the kinetic processes of metal-oxygen reaction throughout the stages of heating the covering, drop growth, and weld pool formation were made by Chen and Kang [210]. A study on modeling the temperature distribution in the electrode for mild steel with three different shielding gases (argon, helium and CO_2) using the PHOENICS computer code and the study of energy balance for heat transfer are presented in [211]. Dixon [212] reports a variety of filler metals to produce ferrite-free welds (affecting magnetic permeability) with low risk of solidification cracking in welded non magnetic steels.

An expert system-based procedure qualified records (PQRs) data base system that combines structural integrity analysis with the AWS D1.1 code for quality welding procedure specifications (WPSs) is given in [213]. In [214], mathematical models were obtained for predicting weld bead geometry and shape relationships for MIG welding of aluminum alloy 5083 using fractional factorial experimental design techniques.

Developments in consumable electrode feeding for robotic GMAW, with a recommendation for push-pull wire feed, were presented in [215]. A theoretical study on the ellipsoidal weld pool during laser welding is presented in [216]. In [217], a study is reported on using GMAW for the collision repair industry to replace the body-over-frame (unibody) car in a quest to reduce the car weight as a means of raising fuel economy.

The development of a dynamic and steady-state model that predicts the electrode extension in the spray mode of GMAW is given in [218, 219]. This study also shows that GMAW acts like a low pass filter for electrode extension with respect to the square of the current (proportional to power). Development of a GMAW process model based on experimentation of three levels (low, medium and high) of power supply voltage, wire feed rate, torch speed, and contact tip-to-workpiece distance (CTWD) is reported in [220]. Using these results, an empirical steady-state model was developed using the statistical package

SYSTAT to get multiple regression linear models from the weld data base. The dynamics were identified by using a step-response test.

A result indicating that pulsed welding current can reduce fume generation rates in GMAW processes compared to steady current can be found in [221]. A comparative study of arc and melting efficiencies of the plasma arc, gas tungsten arc, gas metal arc, and submerged arc welding processes is presented in [222], leading to the result that consumable electrode processes provided the highest efficiencies (0.84), followed by gas tungsten arc (0.67), and finally plasma arc (0.47) processes. A progress report on computational weld mechanics - a structural engineer's point of view of the physics of welding - is given in [223].

Holm [224] developed a method for the establishment of a framework by which a state space model for a whole manufacturing control system such as welding, can be obtained. An experimental and numerical study of power characteristics in GMAW is presented in [225]. This study finds that power input to the arc column increases with both increasing current and arc length and the power input affects melting and tapering of the electrode, size and frequency of droplets, and the solidification of the weld.

In [226], a theoretical model of a GMAW process was developed to predict the anode temperature profile, arc length and arc current, incorporating a one-dimensional thermal model of the moving consumable anode and a two-dimensional model of the arc plasma, with an experimental observation of spray transfer mode for a given welding current and wire feed rate.

Automatic detection of burn-through in GMAW using a parametric model is presented in [227]. This work addresses the problem of automatic detection of burn-through in weld joints. GMAW with pulsed current is used, and welding voltage and current are recorded. As short-circuiting is common between the welding electrode and the workpiece during burn-through, a short-circuit detector is developed to detect these events.

Another work on the effect of pulsed arc on GMAW and the effect of shielding gases is reported in [228]. More experimental works on different aspects of GMAW modeling can be found in [229, 230, 231, 232, 233, 234, 235].

2.13. OTHER ISSUES ON MODELING

Mathematical models relating welding process parameters (such as wire diameter, welding voltage, arc current, welding speed) to the weld bead geometry (weld bead width, depth, and height) were developed with experimental validation of the results in [236].

In [237], the fact that molten metal droplet detachment and plate fusion characteristics are influenced by the parameters of pulsed current is discussed. The complex interdependence of the parameters makes it difficult to select the most suitable combination of parameters for welding. To resolve this problem, a first estimate of the pulse parametric zone based on burnout, droplet detachment, and arc stability criteria was obtained.

Experimentally-observed manifestations of magnetic forces in GMAW are shown and a technique for approximating the temporal evolution of the axial magnetic force from experimentally-measured drop shapes is reported in [238]. A dynamic model of drop detachment in the GMAW is presented in [239] for low and moderate currents in an Argon-rich plasma. The comparisons indicate that the experimental axial magnetic forces are much less potent than the calculated axial magnetic forces when welding-current transients are not present. In [240], an improved fume chamber was constructed, and fume rates were measured with unprecedented precision for both steady and pulsed-current welding of mild steel using 92%, Argon and 8% Carbon Dioxide shielding gas. Comprehensive fume maps were constructed depicting fume rates over a wide range of currents and voltages.

The trends of high-efficiency welding processes in automatic welding systems in Japanese heavy industries are reviewed in [241]. Metal type flux cored wires with CO_2 shielding, high speed rotating GMAW, and mixed MAG have been applied in thick materials with high deposition rates. High current, two-tandem one-side GMAW robots have been developed, with adaptive welding parameter control for change of groove width and joint slope. These results have been applied in curved shell assembly in shipyards.

A study focusing on the properties of SAILMA-450HI plates employing the GMAW process and CO_2 gas is reported in [242]. Implant and elastic restraint cracking tests were conducted to assess the cold cracking resistance of the weld joint under different welding conditions.

In [243], it was argued that twin-wire GMAW systems offer high welding speeds and high depositions rates with a moderate increase in the complexity of the welding system. These systems are now gaining acceptance in arc welding applications requiring high performance.

Narrow-gap gas-shielded metal-arc welding in a vertical-up position and with a horizontal wire feed has the advantage that the heat input is relatively low. Therefore, this process was used for welding tests on 10 mm and 15 mm thick plates made of the fine-grained structural steels S335NL and S460NL as reported in [244]. Various flux-cored wire electrodes served as filler metals.

A study on the welding procedure of gas metal arc welding using cored wire has been reported in [245]. The setting method of the welding condition in GMA welding process with flux cored wire is proposed and its property is investigated.

In [246], it is argued that understanding the mechanisms that contribute to contact tip failure is extremely important to reduce downtime and increase productivity in any GMAW system. The primary function of the contact tip is to effectively transfer the welding current to the consumable electrode as it passes through the center bore and makes electrical contact with the bore surface.

Weld bead characteristics in pulsed GMAW of Al-Mg alloys were investigated in [247] to determine a suitable pulse parameter combination to obtain good-quality welds for a given weld bead size. The effect of pulsed current parameters on GMAW bead geometry was studied, and a single diagram in the non-dimensional form was developed, representing many aspects of weld bead characteristics.

Finite-element prediction of distortion during GMAW using the shrinkage volume approach is presented in [248]. The Shrinkage Volume Method is a linear elastic finite-element modeling technique that has been developed to predict post-weld distortion. An experimental program that investigated the distortion of plain carbon steel plates having different vee-butt preparations carried out verification of the modeled results.

A lumped-parameter, analytical model of material and thermal transfer is established in [249] for metal deposition by a moving, concentrated source. Dynamic description of the distinct width, height, length, and temperature of the ellipsoidal molten puddle is expressed with respect to the torch power, material feed and angle, and the source

2.14. POWER SUPPLIES

motion.

Some basic aspects of geometrical characteristics of pulsed current vertical-up GMAW are discussed in [250]. The performance of pulsed current GMAW in vertical-up position has been studied in reference to thermal behavior of the droplet at the time of deposition affecting the geometrical characteristics of the weld deposit.

2.14 Power Supplies

All welding processes require some form of energy for melting and joining. In general, the welding energy sources are grouped into the following five categories:

1. Electrical sources used for arc welding, resistance welding and electroslag welding.

2. Chemical sources used for oxyfuel gas welding and thermit welding.

3. Focus sources used for lasers, pulsed laser beam welding and continuous wave laser beam welding and electron beam welding.

4. Mechanical sources for friction welding, ultrasonic welding and explosion welding.

5. Other sources for diffusion welding.

The voltage supplied by electric power companies to most industrial operations (120 V, 240 V or 480 V at relatively low currents) is typically too high for usage directly for arc welding is relatively low voltages in the range of 20 to 80 V at very high currents in the range of 30 to 1500 A. Hence, some kind of conversion equipment is required to convert the high voltage to a low voltage. The power required for welding can be either direct current (DC), alternating current (AC) or both DC and AC. The conversion equipment generally used is one or a combination of:

1. A transformer for converting AC to AC.

2. A motor-generator set for converting AC or DC to AC or DC.

3. A solid state convertor such as silicon controlled rectifier (SCR) or thyristor for converting AC to DC.

4. A solid state inverter for converting DC to AC.

A simplified system of classifying various welding power sources is shown in Figure 2.11 [18].

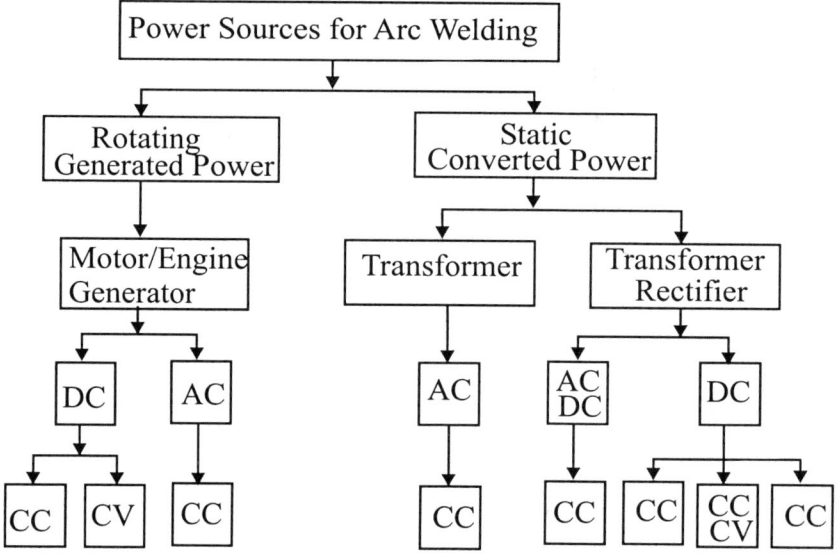

Figure 2.11: Classification of welding power sources.

All the power sources are described by two types of operating characteristics: static and dynamic. A static curve, relating the output-voltage and output-current, is obtained under steady-state conditions using resistive loads. The dynamic curve is determined by measuring the transient variations in output voltage and current. An important characteristic of welding power supplies is the *duty cycle*, defined as the ratio of arc time (the load-on time) to the specified test interval time. Thus, a 60% duty cycle means that 6 out of every 10 minutes, the power source will supply rated current. A 100% duty cycle power supply is designed to output its rated current continuously without exceeding the maximum allowed temperature. For automatic or semiautomatic welding process, the power source should have an 100% duty cycle.

2.14. POWER SUPPLIES

2.14.1 Constant Current (CC)

A constant current (CC) welding machine is one which has a static volt-ampere (V-A) characteristic that tends to produce a nearly constant current. Thus, if the arc length varies due to some external conditions leading to change in arc voltage, the welding current remains substantially constant [14]. In the neighborhood of any operating point, the change in welding current is much less than the corresponding change in load voltage. The no-load or open-circuit voltage is higher than the load voltage depending upon the equivalent resistance of the entire circuit.

A typical volt-ampere (V-A) curve for conventional CC welding power sources is shown in Figure 2.12(a). Note the *drooping* feature

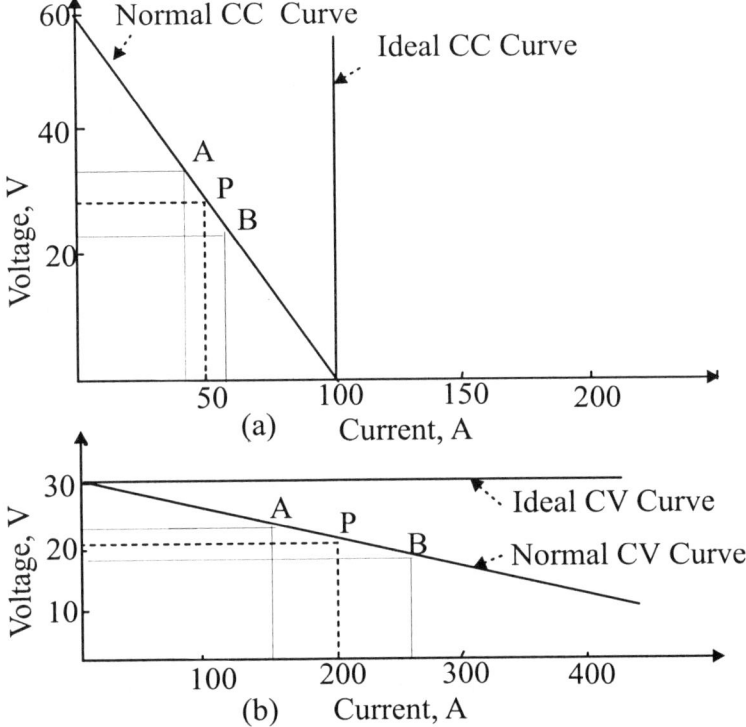

Figure 2.12: Typical volt-ampere curves for: (a) constant current power sources, and (b) constant voltage power sources.

of the V-A curve with negative slope. From the operating point P, a relatively large increase of voltage would result in a relatively small decrease in current. Thus, the higher the negative slope of the V-A curve (or parallel to voltage axis), the better it functions as a constant current source. Therefore, with a consumable electrode welding process, electrode melting rate would be nearly constant even with a small change in arc length.

2.14.2 Constant Voltage (CV)

In a constant voltage (CV) arc welding power source, the static volt-ampere (V-A) characteristic is such that the power source gives a nearly a constant load voltage. A CV source is usually used with a welding machine with a continuously-fed consumable electrode [14].

A typical volt-ampere (V-A) curve for a constant voltage source is shown in Figure 2.12(b). Here, from the operating point P, a large change in current is tolerated with a relatively small change in voltage. This characteristic of a CV power source is suitable for constant feed electrode welding process, such as the GMAW process. A small change in arc length (and hence, arc voltage) will result in a relatively large change in welding current, which will automatically lead to increase or decrease of the electrode melting rate to re-establish the desired arc length (voltage). This phenomena is called *self-regulation*.

2.14.3 Combined CC and CV Power Source

Combination of CC and CV characteristics can be obtained from a single power source by using a variety of electronic and feedback circuits. A typical curve is shown in Figure 2.13.

2.14.4 Pulsed Current

The most common power supplies for GMAW and GTAW are the pulsed current power supplies. Pulsed GMAW power supplies are used to reduce the arc power and wire deposition rates while preserving the desirable spray transfer mode. For more details of the pulsed current operation see the previous section on Metal Transfer Characteristics.

2.15. OTHER ISSUES ON POWER SUPPLIES

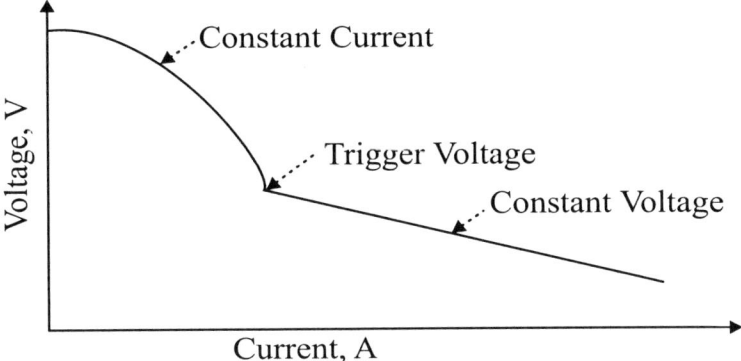

Figure 2.13: Typical volt-ampere curves for a combined CC and CV power source.

2.14.5 Inverters

With the introduction of microprocessors, the primary power sources have become inverters using silicon-controlled rectifiers. A class-H power supply that was more efficient than the conventional class-A power source was built using switching transistors and diodes [185]. This power source delivers up to 45 volts and 500 amperes with a frequency response up to several decades of kHz. Power supplies using solid-state technology and power-inverter technology and capable of being easily programmable and delivering controlled arcs are discussed in [251, 252].

A report [253] discusses the choice of inverters as ideal power sources for welding due to the fact that increasing the switching frequencies decreases the size of the welding power source and produces a smoother output.

2.15 Other Issues on Power Supplies

In [254], for GMAW during the short-circuiting welding mode, a current programmer was designed in conjunction with a high performance solid-state power source operating in the constant current mode. The programmer allows precise control of output current during the short-circuiting and arcing phases of welding. Details about power sources for welding and for developing a new power source with new current waveforms, that offers spatter suppression, stability at high welding speeds, and improved joint tracking in automatic applications of welding are discussed in [255].

2.16 Classification of References by Section

Here, we provide a table containing the various references according to each section of this chapter. This will provide a ready reference to the interested reader to search for relevant references in each section.

Table 2.1: Section by Section List of References

Section	Reference Numbers
2.1 Gas Metal Arc Welding	[13]-[19]
2.2 Physics of Welding	[14],[15],[20]-[66]
2.3 Melting Rate	[67]-[72]
2.4 Metal Transfer Characteristics	[14],[67],[73]-[117]
2.5 Weld Pool	[32],[37],[63],[118]-[146]
2.6 Process Voltages	[117],[147]
2.7 Heat and Mass Transfer	[148],[149]
2.8 Process Variables	[150],[159]
2.9 INEEL/ISU Model	[101],[108],[109],[117],[149],[160]-[171]
2.10 Empirical and Statistical Model	[172]-[181]
2.11 Modeling by System Id. and Est.	[182],[183]
2.12 Distributed Parameter Model	[119],[184]
2.13 Intelligent Modeling	[157],[185]-[196]
2.14 Other Issues on Modeling	[172],[197]-[250]
2.15 Power Supplies	[14],[18],[185],[251]-[253]
2.16 Other Issues on Power Supplies	[254],[255]

References List for Chapter 2

[13] D. E. Hardt. Modeling and control of manufacturing processes: getting more involved. Transactions of the ASME, Journal of Dynamic Systems, Measurement and control. 115, Pages 291-300, June, 1993.

[14] R. L. O'Brien. Welding handBook: Welding processes. Volume 2, Eighth Edition. Maimi, FL, American Welding Society, 1991.

[15] L. P. Connor. Welding HandBook. Eigth Edition, Volume 1. Miami, FL, American Welding Society, 1987.

[16] J. G. Bollinger. Control of Welding Processes. National Research Council. Washington, DC, 1987.

[17] G. Begin and J. P. Boillot. Welding adaptive functions performed through infrared(IR) simplified vision schemes", Proceedings of the 3rd International conference on Robot Vision and Sensory Controls RoViSeC3. Cambridge, MA, 449, Pages 328-337. November, 1983.

[18] H. B. Cary. Modern Welding Technology. Prentice Hall, Englewood Cliffs, NJ, Second Edition, 1989.

[19] G.- D. Wu and R. W. Richardson. The dynamic response of self-regulation of the welding arc. Proceedings of the ASM 2nd International Conference on Recent Trends in Welding Research. Gatlinburg, TN, May, S. A. David and J. M. Vitek, Pages 929-933, 1989.

[20] M. H. Hoyaux. Arc Physics. Springer-Verlag, New York, 1968.

[21] S. S. Glickstein. Arc modeling for welding analysis. Proceedings of the International Conference on on Arc physics and Weld Pool Behavior. London, UK, May, Pages 1-16, 1979.

[22] S. S. Glickstein. Basic studies of the arc welding process. Trend in Welding Research in The Unite States. American Society for Metals. S. A. David, Mata park, OH. Pages 3-49, 1982.

[23] S. S. Glickstein and E. Friedman. Weld modeling applications. Welding Journal, Volume 63, Pages 38-42. September, 1984.

[24] C. E. Jackson. The science of arc welding: Part I-definition of arc. Welding Journal. Volume30, Pages129s-140s, April, 1960.

[25] C. E. Jackson. The science of arc welding Part II-consumable-electrode welding arc. Welding Journal, Pages 177s-190s. May, 1960.

[26] C. E. Jackson. The science of arc welding Part III-what the arc does", Welding Journal, Pages 225s-230s. May, 1960.

[27] Dirk Dzelnitzki. Increasing the deposition volume or the welding speed - advantages of heavy duty MAG welding. Welding Research Abroad, Volume 45(3),Pages 10-17, 1999.

[28] C. D. Yoo, Y. S. Yoo and H. K. Sunwoo. Investigation on arc light intensity in gas metal arc welding. Part1 : Relationship between arc light intensity and arc length. Proceedings of the Institution of Mechanical Engineers. Pages 345-353,London, England, 1997.

[29] E. Ide and M. Matsui and M. Matsumoto and T. Kawano and H. Iwabuchi and J. Wakiyama. Development of Arc Simulation system on gas shielded metal arc welding process. Technical Review - Mitubishi Heavy Industries. Volume 34(1), Pages 1-4, 1997.

[30] D. Farson, C. Conrady, J. Talkington, K. Baker, Kerschbaumer and P. Edwards. Arc Initiation in gas metal arc welding. Welding Journal,Volume 77 (8), Pages 315s-321s. August, 1998.

REFERENCES LIST FOR CHAPTER 2

[31] N. Christensen, V.L. Davies and K. Gjermundsen. Distribution of temperatures in arc welding. British Welding Journal, Volume 54(2),Pages 54-75. February, 1965.

[32] D. Rosenthal. Mathematical theory of heat distribution during welding and cutting. Welding Journal, Volume 20(5), 220s-234s, May, 1941.

[33] P. Jhaveri, W. G. Moffatt and C. M. Adams Jr. The effect of plate thickness and radiation on heat flow in welding and cutting. Welding Journal, Volume 41(1), Pages 12s-16s. January, 1962.

[34] W. G. Essers and R. Walter. Heat transfer and penetration mechanisms with GMA and plasma-GMA welding. Welding Journal, Volume 60, Pages 37s-42s. February, 1981.

[35] J. H. Waszink and G. J. P. M. van den Heuvel. Heat generation and heat flow in the filler metal in GMA welding. Welding Journal, Volume 61, Pages 269s-282s. August, 1982.

[36] C. L. Tsai. Modeling of thermal behaviors of metals during welding. Trends In Welding Research in the United States. American Society for Metals. S. A. David. Pages 91–108, 1982

[37] C. L. Tsai. Heat Distribution and Weld Bead Geometry in Arc welding. MIT, Cambridge, MA, 1983.

[38] M. C. Tsai and S. Kou. Electromagnetic -force- induced convection in weld pools with a free surface. Welding Journal, Pages 241s-246s, June, 1990.

[39] H. D. Solomon and S. Levy. HAZ temperatures and cooling rate as determined by a simple computer program. Trends In Welding Research In The United States. American Society for Metals. S. A. David, Metals Park, OH, Pages 173–205, 1982.

[40] J. H. Waszink and M. J. Piena. Experimental investigation of drop detachment and drop velocity in GMAW. Welding Journal, Pages 289s-298s, November, 1986.

[41] O. W. Blodgett. Calculating cooling rates of arc spot welds. Welding Journal, Pages 17-30, December, 1987.

[42] U. Dilthey and R. Killing. Contribution to calculating the heat input in pulsed arc gas shielded metal arc welding. Schweissen - Schneiden, Volume 39(10), Pages E160-E162", German, October, 1987.

[43] J.V. Beck. Inverse problems in heat transfer with applications to solidification and welding. Proceedings of the 5th International Conference on Modelling of Casting and Welding Processes. Davos, Switzerland. M. Rappaz. Pages 503-515, September, 1990.

[44] J. W. Kim and S. J. Na. A study on the three dimensional analysis of heat and fluid flow in gas metal welding using boundary fitted coordinates. Proceedings of Winter Annual meeting of the American society of Mechanical Engineers. ASME, Atlanta, GA, Pages 159-173, 1991.

[45] Y. S. kim, D. M. McEligot and T. W. Eagar. Analysis of electrode heat transfer in gas metal arc welding. Welding Journal, Pages 20s, January, 1991.

[46] A. Lesnewich. Technical commentary: Observations regarding electrical current flow in the gas metal arc. Welding Journal, Pages 171s-172s, July, 1991.

[47] A. Matsunawa. Modeling of heat and fluid flow in arc welding. Proceedings of the ASM 3rd International Conference on Trends in Welding Research. S.A. David and J. M. Vitek. Gatlinburg, TN, June 1-5, Pages 3-16, 1992.

[48] A. R. Ortega, L. A. Bertram, E.A. Fuchs and K. W. Mahin and D. V. Nelson. Thermomechanical modeling of a stationary gas metal arc weld: a comparison between numerical and experimental results. Proceedings of the ASM 3rd International Conference on Trends in Welding Research. Gatlinburg, TN, June, S. A. David and J. M. Vitek. Pages 89-93, 1992.

[49] M. Hang and A. Okada Computation of GMAW welding heat transfer with boundary element method. Advances in Engineering Software, Volume 16, Pages 1-5, 1993.

[50] S. Kumar and S. C. Bhaduri. Three dimensional finite element modeling of gas metal arc welding. Metallurgical and Materials Transactions B: Process Metallurgy and Materials, Volume 25(3), Pages 435-441, June, 1994.

[51] S. Kumar and S. C. Bhaduri. Theoretical investigation of penetration characteristics in gas metal-arc welding finite element method. Metallurgical and materials Transactions B: Process Metallurgy and Material. Volume 26B(3), Pages 611-624. June, 1995.

[52] M. Ushio and C. S. Wu. Mathematical modeling of three-dimensional heat and fluid in a moving gas metal arc weld pool", Metallurgical and Materials Transactions. B: Process Metallurgy and Materials Processing Science", Volume 28(3),Pages 509-516, June, 1997.

[53] I. S. Kim and A. Basu. Mathematical Model of heat transfer and fluid flow in the gas metal arc welding process", Journal of Materials Processing Technology, Volume 77(1-3),Pages 17-24, May,1998.

[54] M. A. Wahab, M. J. Painter and M. H. Davies. Prediction of the temperature distribution and weld pool geometry in the gas metal arc welding process. Journal of Materials Processing Technology. Vol 77, No 1-3, Pages 233-239, May, 1998.

[55] C. S. Wu and J. S. Sun. Modelling the arc heat flux distribution in GMA welding. Computational Materials Science,Volume 9(3-4), Pages 397-402. January,1998.

[56] H. Schrier, M. Sutton, Y. Chao, H. Bruck and J. Dydo. Full-Field temperature and three-dimensional displacement measurements in hostile environments. Technical Paper Society of Manufacturing Engineers. MR99-140, Pages 1-5,March,1999.

[57] Y. J. Chao and Q. Xinhai. Three-dimensional modeling of gas metal arc welding process. Technical Paper - Society of Manufacturing engineers. MR99-164,Pages 1-6, March,1999.

[58] Z. Yang and T. Debroy. Modeling macro-microstructures of gas-metal-arc welded HSLA-100 steel. Metallurgical and Materials

Transactions A: Physical Metallurgy and Materials Science. Volume 30(3),Pages 483-493,Year 1999.

[59] C. Limmaneevichitr and S. Kou. Experiments to simulate effect of Marangoni convection on weld pool shape. Welding Journal (Miami, FL). Volume 79(8),Pages 231s-237s. August, 2000.

[60] K. E. Dorschu. Control of cooling rates in steel weld metal. Welding Journal,Volume 47(2),Pages 49s-62s, February,1968.

[61] J. G. Garland. Weld pool solidification control. Metal Construction and British Welding Journal. Volume 6(4),Pages 121-127, April, 1974.

[62] K. Ishizaki. Solidification of the molten metal pool and bead formation. Proceedings of an International Conference on Arc Physics and Weld Pool Behavior", London, UK, May, Pages 267-277,1979.

[63] A. J. Aendenroomer. Weld Pool Oscillation for Penetration Sensing and Control. Technical School of Delft University, Delft, The Netherlands,1996.

[64] E. Craig. The plasma arc process-a review. Welding Journal, Pages 19-25. February,1988.

[65] W. G. Essers. Plasma with GMA welding. Welding Journal,Volume 55, May,Pages 394-400,1976.

[66] T. W. Eager. The physics and chemistry of welding processes In Advances in Welding Science and Technology,S. A. David. Metals Park, OH. Pages 291-298,1986.

[67] J. Norrish. Advanced welding processes. Institute of Physics Publishing. Bristol, UK, 1992.

[68] A. Lesnewich. Control of melting rate and metal transfer in gas -shielded metal-arc welding part 1: Control of electrode melting rate. Welding Journal, Volume 37,(8),Pages 343s-353. August,1958.

[69] R. S. Chandel. Mathematical modeling of melting rates for submerged arc welding. Welding Journal, Pages 135s-140s. May,1987.

REFERENCES LIST FOR CHAPTER 2

[70] E. Halmoy. Wire melting rate, droplet temperature, and effective anode melting potential. Proceedings of an International Conference on Arc Physics and Weld Pool Behavior, London, UK. The Welding Institute, Pages 49-57, 1979.

[71] J. Tusek. Mathematical model for the melting rate in welding with a multiple-wire electrode. Journal of Physics D: Applied Physics, Volume 32(14),Pages 1739-1744,1999.

[72] Z. Bingul and G. Cook. Dynamic modeling of GMAW process. Proceedings - IEEE International Conference on Robotics and Automation. Piscataway, New Jersey,Pages 3059-3064,1999.

[73] R. Kovacevic, Y. M. Zhang, E. Liguo and H. E. Beardsley. Dynamics of droplet geometry during metal transfer in GMAW - a model for process control. ASME International Mechanical Engineering Congress and Exposition. Atlanta, Georgia, November, Pages 143-144, 1996.

[74] L. Quintino, R. Riberio and J. Faira. Classification of GMAW transfer using neural networks. International Journal for the Joining of Materials. Volume 11(1), Pages 6-8, 1999.

[75] S. K. Choi, Y. S. Kim and C. D. Yoo. Dimensional analysis of metal transfer in GMA welding. Journal of Physics D: Applied Physics. Volume 32(3),Pages 326-334, February, 1999.

[76] W. H. Chen, B. A. Chin and S. Nagarajan. Infrared sensing for adaptive arc welding. Welding Journal,Pages 462s, November, 1989.

[77] H. F. Fan and R. Kovacevic. Dynamic analysis of globular metal transfer in gas metal arc welding - a comparison of numerical and experimental results. Journal Of Physics D: Applied Physics. Volume 31(20), Pages 2929-2941, October, 1998.

[78] J. Haidar. Analysis of the formation of metal droplets in arc welding. Journal of Physics D: Applied Physics. Volume 31(10), Pages 1233-1244, May, 1998.

[79] G. N. Metko. Pulsed spray transfer welding", Proceedings of Conference on Sheet Metal Welding Conference The Latest Technology for Sheet Metal Joining. Detroit, MI, American Welding Society, 1984.

[80] D. S. Mathews. GMAW Short Circuiting Transfer: Stabilities, Instabilities, and Joule Heating Model. Vanderbilt University. Nashville, TN, 1988.

[81] K. Andersen, G. E. Cook, L. Yizhang, D. S. Mathews and M. D. Randall. Modeling and control parameters for GMAW: short-circuiting transfer. Advances in Manufacturing Systems: Integration and Processes. D. A. Dorfield, Dearborn, MI, Pages 413-421, 1989.

[82] G. E. Cook D. R. DeLapp, R. J. Barnett and A. M. Strauss. Modeling and control parameters for GMAW, short- circuiting transfer. Proceedings of the ASM 4th International Conference on Trends in Welding Research. Gatlinburg, TN. H. B. Smartt, J. A. Johnson and S. A. David. June 5-8, Pages 721-726, Year 1995.

[83] A. Lesnewich. Control of melting rate and metal transfer in gas-shielded metal-arc welding part II - control of metal transfer. Welding Journal, Pages 418s-425s. September, 1958.

[84] W. Spraragen and B. A. Lengyel. Physics of the arc and the transfer of metal in arc welding: A review of the literature to February 1942", Welding Journal, Volume 1, Pages 2s–42s. January, 1943.

[85] J. A. Hirschfield. Welding aluminum, more on gas metal-arc welding. Welding Journal, Volume 54, Pages 28-30. January, 1975.

[86] G. J. Ogilvie and I. M. Ogilvy. The pulsed GMA process in automatic welding. Proceedings of 31st Annual Conference of The Australian Welding Institute, Pages 16-19 Sydney, Australia, October, 1983.

[87] K. S. Ogilvie. Modeling of the gas metal arc welding process for control of arc length, Dissertation. University of Michigan, Ann Arbor, MI, 1991.

[88] L. E. Cram. A numerical model of droplet formation. Proceedings of the 1983 International Conference on Computational Techniques and Applications. J. Noyce and C. Fletcher. Sydney, Australia, Elsevier Science Publishers, 1984.

[89] L. Quintino and C. J. Allum. Pulsed GMAW: interactions between process parameters - Part 1. Welding and Metal Fabrication, Pages 85-89, March, 1984.

[90] L. Quintino and C. J. Allum. Pulsed GMAW: interactions between process parameters - Part 2, Welding and Metal Fabrication, Pages 126-129, April, 1984.

[91] W. G. Essers and M. R. M. Van Gompel. Arc control with pulsed GMA welding, Welding Journal, Pages 26-32, June, 1984.

[92] E. M. Trindade and C. J. Allum. Characteristics in steady and pulsed current GMAW. Welding and Metal Fabrication, Pages 264-272, 1984.

[93] S. Ueguri, K. Hara and H. Komura. Study of metal transfer in pulsed GMA welding. Welding Journal, Pages 242s-250s, August, 1985.

[94] H. R. Saedi and W. Unkel. Arc and weld pool behavior for pulsed current GTAW. Welding Journal, Volume 67(11), Pages 247s-255s, 1988.

[95] D. E. Clark, C. Buhrmaster and H. B. Smartt. Droplet transfer mechanisms in GMAW. Proc. of 2nd Intl. Conf. on Trends in Welding Research. Gatlinburg, TN, June, 1989.

[96] S. Liu and T. A. Siewert. Metal transfer in gas metal arc welding: droplet rate. Welding Journal, Volume 68(2), Pages 52s-58s, February, 1989.

[97] M. J. Lu and S. Kou. Power inputs in gas metal arc welding of aluminum-part 1. Welding Journal, Pages 382s-388s. September, 1989.

[98] M. J. Lu and S. Kou. Power inputs in gas metal arc welding of aluminum-part 2. Welding Journal, Pages 452s-456s, November, 1989.

[99] S. Brandi and C. Taniguchi and S. Liu. Analysis of metal transfer in shielded metal arc welding Welding Journal, Pages 261s-270s. October, 1991.

[100] Y. S. Kim and T. W. Eagar. Analysis of metal transfer in gas metal arc welding. Welding Journal, Volume 72(6), Pages 269s-278s, June, 1993.

[101] E. W. Reutzel, C.J. Einerson, J. A. Johnson, H. B. Smartt, T. Harmer and K. L. Moore. Derivation and calibration of a gas metal arc welding GMAWdynamic droplet model. Proceedings of the ASM 4rd International Conference on Trends in Welding Research, Pages 377-384. H. B. Smartt and J. A. Johnson and S. A. David. Gatlinburg, TN, June, 1995.

[102] J. Haidar and J. J. Lowke. Predictions of metal droplet formation in arc welding. Journal of Physics D: Applied Physics, Volume 29(12), Pages 2951-2960. December, 1996.

[103] R. Kovacevic, Y. M. Zhang, E. Liguo and H. Beardsley. Dynamic analysis metal transfer process for GMAW control. ASM Journal of Engineering Materials and Technology. July, 1996.

[104] Anonymous. Gas metal arc welding: transfer modes. Welding Journal, Volume 76, Pages 58-59, February, 1997.

[105] J. A. Johnson and N. M. Carlson. Noncontact ultrasonic sensing of weld pools for automated welding. Proceedings of 3rd International Symp. on Nondestructive Characterization of Materials", Saarbrucken,FRG, Pages 854-861, 1988.

[106] J. A. Johnson, H. B. Smartt, D. E. Clark, N. M. Carlson, A. D. Watkins and B. J. Lathcoe. The dynamics of droplet formation and detachment in gas metal arc welding. Proceedings of the Fifth International Conference on Modeling of Casting and Welding Processes. Davos, Switzerland. September, Pages 139-146, 1990.

REFERENCES LIST FOR CHAPTER 2

[107] J. A. Johnson, N. M. Carlson, H. B. Smartt and D. E. Clark. Process control of GMAW: Sensing of metal transfer Mode. Welding Journal, Volume 70, Pages 91s-99s. April, 1991.

[108] J. A. Johnson, M. Waddoups, H. B. Smartt and N. M. Carlson Dynamics of droplet detachment in GMAW Proceedings of the 3rd International Conference on Trends in Welding Research, Pages 987-997. S. A. David and J. M. Vitek, Gatlinburg, TN. June, 1992.

[109] A. D. Watkins, H. B. Smartt and J. A. Johnson. A dynamic model of droplet growth and detachment in GMAW. Proceedings of ASM 3rd International Conference on Trends in Welding Research. Pages 993-1002. S. A. David and J. M. Vitek, Gatlinburg, TN. June, 1992.

[110] L. A. Jones, T. W. Eagar and J. H. Lang. Investigations of drop Detachment Control in Gas Metal Arc Welding. Proceedings of the 3rd International Conference on Trends in Welding Research, Pages 1009-1013. S. A. David and J. M. Vitek,Gatlinburg, TN, June, 1992.

[111] G. Adam and T. A. Siewert. On-line arc welding data acquisition and analysis system. Proceedings of the 2nd International Conference on Trends in Welding Research, Pages 979-983. Gatlinburg, TN, May, 1989.

[112] G. Adam and T. A. Siewert. Sensing of GMAW droplet transfer modes using an ER 100s-1 electrode. Welding Journal, Volume 69(3), Pages 103s-108s. March, 1990.

[113] S. Subramaniam, D. R. White, D. J. Scholl and W. H. Weber . In situ optical measurements of liquid drop tension in gas metal arc welding. Journal of Physics D: Applied Physics, Volume 31(16), 1963-1967. August, 1998.

[114] P. K. Ghosh, S. R. Gupta and H. S. Randhawa. Characteristics and critically of bead on plate deposition in pulsed current vertical-up GMAW of steel. International Journal for the Joining Materials, Volume 11(4), Pages 99-110., 1999.

[115] Tomasz Wegrzyn. Classification of metal weld deposits in terms of the amount of Oxygen. Proceedings of the International Offshore and Polar Engineering Conference, Pages 212-216. Golden, Colorado, 1999.

[116] M. H. Davies, M. Wahab and M. J. Painter. Investigation of the interaction of a molten droplet with a liquid weld pool surface: a computational and experimental approach. Welding Journal (Miami, FL), Volume 79(1),Pages 18s-23s. January, 2000.

[117] J. F. Lancaster. The Physics of Welding. Pergamon Press, Second Edition. Oxford, UK, 1986.

[118] R. T. Choo, J. Szekely and R. C. Westoff. Modeling of high-current arcs with emphasis on free surface phenomena in the weld Pool. Welding Journal, Pages 346s-361s, September, 1990.

[119] C. C. Doumanidis. Multiplexed and distributed control of automated welding. IEEE Control Systems, Volume 14(4), Pages 13-24. August, 1994.

[120] D. Rosenthal. The theory of moving surfaces of heat and its application to metal treatment. Transactions of ASME, Pages 849-866. November, 1946.

[121] B. E. Bates and D. E. Hardt. A real-time calibrated thermal model for closed-loop weld bead geometry control. ASME Journal of Dynamic Systems, Measurement and Control, Volume 107,Pages 25-33. March, 1985.

[122] R. J. Barnett. Sensor development for multi-parameter control of gas tungsten arc welding. Vanderbilt University, Nashville, TN, 1993.

[123] K. Andersen. Synchronous weld pool oscillation for monitoring and control. Vanderbilt University, Nashville, TN, 1993.

[124] K. Andersen, G. E. Cook, R. J. Barnett and A. M. Strauss. Synchronous weld pool oscillation for monitoring and Control. IEEE Transactions on Industry Applications, Volume 33(2), Pages 464-471. March/April, 1997.

[125] D. E. Hardt. Measuring weld pool geometry from pool dynamics. Proceedings of 3rd Conference on Modeling of Casting and Welding Process, Pages 3-17, 1988.

[126] R. T. Deam. Weld pool frequency: A new way to define a weld procedure. Proceedings of 2nd International Conference on Trends in Welding Research. S. A. David and J. M. Vitek, Gatlinburg, TN. Pages 967-971, May 18-22, 1989.

[127] Y. H. Xiao and G. den Quden. A study of GTA weld pool oscillation. Welding Journal, Volume 69(8), Pages 289s-293s,1990.

[128] C. D. Yoo. Effects of weld pool condition on pool oscillation. The Ohio State University, 1990.

[129] I. S. Kim, A. Basu and E. Siores. Mathematical models for control of weld beam penetration in the GMAW process. International Journal Of Advanced Manufacturing Technology. Volume 12(6), Pages 393-401,1996.

[130] L. Feng, S. Chen, L. Liangyu and S. Li. Static Equilibrium model for the bead formation in high speed gas metal arc welding", China Welding (English Edition), Volume 7(1), Pages 22-27. December, 1998.

[131] S. Ohring and H. J. Lugt. Numerical simulation of a time-dependent 3-D GMA weld pool due to a moving arc. Welding Journal (Miami, FL), Volume 78(12), Pages 416s-424s, 1999.

[132] J. S. Sun, C. S. Wu and J. Q. Gao. Modeling the weld pool behaviors in GMA welding. International Journal for the Joining Materials, Volume 11(4), Pages 112-117, 1999.

[133] D. E. Hardt, D. A.Garlow and J. B. Weinert. A model of full penetration arc welding for control system design. Proceedings of the Winter Annual Meeting of the ASME on Control of Manufacturing Processes and Robotic Systems. D. E. Heart and W. J. Book. ASME, New York, Pages 121-135, November, 1983.

[134] D. E. Hardt, D. A. Garlow and J. B. Weinert. A model of full penetration arc welding for control system design. Transactions of

the ASME, Journal of Dynamic systems, Measurement and Control. Volume 107, Pages 40-46, March, 1985.

[135] S. Kou and Y. Le. Improving weld quality by low frequency arc oscillation. Welding Journal, Pages 51-55. March, 1985.

[136] M. Lin and T. W. Eagar. Influence of arc pressure on weld pool geometry. Welding Journal, Volume 64, Pages 163s-169s, June,1985.

[137] A. S. Tam and D. E. Hardt. Weld pool impedance for pool geometry measurement: stationary and nonstationary pools. ASME Transactions: Journal of Dynamic systems, Measurement and Control. Volume 111, Pages 545–553, December, 1989.

[138] M. Zacksenhouse and D. E. Hardt. Weld pool impedance identification for size measurement and control. ASME Trans.: J. of Dynamic System, Measurement, and Control. Vol 105, No 9, Pages 179-184. September, 1983.

[139] C. D. Sorensen and T. W. Eagar. Modeling of oscillation in partially penetrated weld pools. Transactions of the ASME, Journal of Dynamic Systems, Measurement and control. Volume 112(3), Pages 469–474, September, 1990.

[140] C. C. Doumanidis. GMAW weld bead geometry: A lumped dynamic model. Proceedings of the 3rd International Conference on Trends in Welding Research. S. A. David and J. M. Vitek, Gatlinburg, TN, June, Pages 63-67, 1992.

[141] C. C. Doumanidis. Hybrid modeling for control of weld dimensions. Japan/USA Symposium on Flexible Automation, Pages 317-323, 1992.

[142] C. D. Yoo and R. W. Richardson. An experimental study on sensitivity and signal characteristics of weld pool and oscillation. Transaction of Japan Welding Society, Volume 24, Pages 54-62. October, 1993.

[143] J. W. Kim and S. J. Na. A study on the effect of contact tube to workpiece distance on weld pool shape in gas metal arc welding.

Welding Journal, Volume 74(5), Pages 141-152. Miami, FL, May 1995.

[144] J. W. Kim and S. J. Na. A Study on prediction of welding current in gas metal arc welding Part 1: Modeling of welding current in response to change of tip-to-workpiece distance. Proceedings of the Institution of Mechanical Engineers,Part B: Journal of Engineering Manufacture, Volume 205, Pages 59-63, 1991.

[145] J. W. Kim and S. J. Na. A study on prediction of welding current in gas metal arc welding Part 2: Experimental modeling of relationship between welding current and tip-to-workpiece distance and its application to weld seam tracking system. Proceedings of the Institution of Mechanical Engineers Part B: Journal of Engineering Manufacture. Volume 205, Pages 65-69, 1991.

[146] N. Zhen C. and P. Dong. Modeling of GMA weld pools with consideration of droplet impact. Journal of Engineering Materials and Technology, Transactions of the ASME. Volume 120(4), Pages 313-320, October, 1998.

[147] R. B. Madigan. Control of gas metal arc welding using arc light sensing. Colorado School of Mines. Golden, CO, 1994.

[148] H. B. Smartt and C. J. Einerson. A model for heat and mass input control in gas metal arc welding. Welding Journal, Volume 72(5), Pages 217s–229s. May, 1993.

[149] S. Ozcelik, K. L. Moore and D. S. Naidu. Adaptive control of a gas metal arc welding (GMAW) process. Measurement and Control Engineering Research Center, Idaho State University. Pocatello, ID, April, 1997.

[150] D. A. Dornfeld, M. Tomizuka and G. Langeri. Modeling and adaptive control of arc welding processes. Measurement, Control in Batch Manufacturing. ASME, D. E. Hardt. New York, NY, Pages 53-64, November, 1982.

[151] C. M. Payares, M. Dorta and P. E. Munoz. Influence of the welding variables on the mechanical properties in butt joints for

aluminum 6063-T5. American Society of Mechanical Engineers. Pressure Vessels and Piping Division. Volume 393, Pages 339-344, 1999.

[152] J. H. Waszink and G. J. P. M. van den Heuvel. Measurements and calculations of the resistance of the wire extension in arc welding. Proceedings of the International Conference on Arc Physics and Weld pool Behavior. London, UK, May, Pages 227-239,1979.

[153] V. R. Dillenbeck and L. Castagno. The effects of various shielding gases and associated mixtures in GMA welding of mild steel", Welding Journal, Volume 66, September,Pages 45-49, 1987.

[154] N. Stenbacka and O. Svensson. Some observations on pore formation in gas metal arc welding. Scandinavian journal of Metallurgy,Volume 16(4), Pages 151–153, 1987.

[155] R. E. Francis, J. E. Jones and D. L. Olson. Effect of shielding gas oxygen activity on weld metal microstructure of GMA welded microalloyed HSLA steel. Welding Journal, Pages 408s-415s, November, 1990.

[156] G. E. Cook. Feedback and adaptive control in automated arc welding systems. Metal Construction, Pages 551-556, Volume 13(9). September, 1981.

[157] G. C. Cook, K. Anderson and R. J. Barrett. Feedback and adaptive control in welding. Proceedings of the 2nd International Conference on Trends in Welding Research. Gatlinburg, TN, May, 1989 S. A. David and J. M. Vitek, Key Note Address, Pages 891-903, 1989.

[158] G. E. Cook, K. Andersen and R. J. Barrett. Welding and bonding. The Electrical Engineering Hand Book. CRC Press, Boca Raton, FL. R. C. Dorf, Pages 2223-2237, 1993.

[159] H. C. Wezenbeek. A System for Measurement and Control of Weld Pool Geometry in Automatic Arc Welding. Technische Univ. Eindhoven Netherlands, Dept. of Electrical Engineering. Eindhoven, The Netherlands, 1992.

REFERENCES LIST FOR CHAPTER 2

[160] K.L. Moore, D.S. Naidu, M. Abdelrahman and A. Yesildirek. Advanced Welding Control Project: Annual Report FY96. Measurement and Control Engineering Research Center. Idaho State University, Pocatello, ID. 28th June, 1996.

[161] K.L. Moore, D.S. Naidu, S. Ozcelik, R. Yender and J. Tyler. Advanced Welding Control Project: Annual Report, FY97. Idaho State University, Pocatello, ID, July, 1997.

[162] M. E. Shepard and G. E. Cook. A frequency-domain model of self-regulation in gas metal arc welding. Proceedings of the ASM 3rd International Conference on Trends in Welding Research. Gatlinburg, TN, S. A. David and J. M. Vitek. June, Pages 899-903, 1992.

[163] T. M. Harmer, K. L. Moore, H. B. Smartt, J. A. Johnson and E. R. Reutzel. Modeling and simulation of a GMAW, Unpublished Work, 1994.

[164] J. H. Waszink and L. H. Graat. Experimental investigation of the forces acting on a drop of weld metal. Welding Journal, Volume 62, April, Pages 108s-116s, 1983.

[165] R. Shaw. The Dripping Faucet as a Model Chaotic System. Ariel Pess, Inc. Santa Cruz, CA, 1984.

[166] H. Fujimura and E. Ide and H. Inoue. Joint tracking control sensor of GMAW: Development of method and equipment for position sensing in welding with electric arc signals (Report 1). Transactions of the Japan welding Society, Volume 18(1), Pages 32-40. April, 1987.

[167] H. Fujimura and E. Ide and H. Inoue. Weave amplitude control sensor of GMAW: Development of method and equipment for position sensing in welding with electric arc signals (Report 2). Transactions of the Japan Welding Society, Volume 18(1), Pages 41-45. April, 1987.

[168] H. E. Fujimura, E. Ide and H. Inoue. Estimation of contact tip-workpiece distance in gas metal arc welding. Welding International, Volume 2(6), Pages 522-528, 1988.

[169] R.M. Schulkes. The evolution and bifurcation of a pendant drop. Journal of Fluid Mechanics, Volume 278, Pages 83–100, 1994.

[170] M. Waddoups. Detection and Modeling of Droplet Detachment in Gas Metal Arc Welding. Idaho State University. Pocatello, ID, 1994.

[171] B. W. Greene. Arc current control of a robotic welding system: modeling and control system design. Illinois Univ. at Urbana-Champaign, Coordinated science Lab. Champaign, IL. DC-114-UILU-ENG-89-2227, 1990.

[172] J. C. McGlone, J. Doherty and S. B. Jones. Developments in arc welding procedure selection. Proceedings of ANS Conference on Welding and Fabrication in the Nuclear Industry. Nuclear Energy Society. London, UK. Pages 283-289, April, 1979.

[173] J. J. Hunter, G. W. Bryce and J. Doherty", On-line control of the arc welding process. Proc. of the Int. Conf. on Developments in Mechanized Automated and Robotic Welding Process. The Welding Institute, London, UK. Pages P37-1-P37-11, November, 1980.

[174] V. A. Parshin, A. K. Seliverstov, A. V. Parfenova, S. B. Shakhanov and V. I. Shakhvatov", Algorithm for the control of the dimensions of welded joints. Welding Production (GB). Volume 28(10), Pages 3-5. October, 1981.

[175] Y. Arata, K. Inoue, Y. Shibata, M. Tamaoki and H.Akashi Automatic control of arc welding (Report I)- Algorithm for automatic selection of optimum welding condition. Transactions of Japan Welding Research Institute (JWRI), Volume 8(1), Pages 1-11, 1979.

[176] G. I. Segatskii, S. V. Dubovetskii and O. G. Kasatkin. Models for open control of CO_2 weld formation. Automatic Welding(GB). Volume 36(2), Pages 18–21. February, 1983.

[177] J. Doherty, S. J. Holder and R. Baker. Computerized guidance and process control. Proceedings of Third International Confer-

REFERENCES LIST FOR CHAPTER 2   85

ence on Robot Vision and Sensory Controls RoViseC3", Cambridge, MA. D. P. Cassasent and E. L. Hall. Volume 449, Pages 482-487. November, 1983.

[178] M. Galopin and E. Boridy. Une approche statistique du choix d'un mode operatoire de soudage. Soudage et Techniques Connexe. Volume 37(11-12), Pages 403-412, 1983.

[179] S.V. Dubovetskii. The application of statistical models of weld formation to the design and control of automatic arc welding conditions. Proceedings of the EWI 1st International Conference on Advanced Welding System. London, UK. P. T. Houldcroft, Pages210-236", November, 1985.

[180] R. T. Deam, S. W. Simpson and J. Haidar. Semi-empirical model of the fume formation from gas metal arc welding. Journal of Physics D: Applied Physics. Volume 33(11), Pages 1393-1402, 2000.

[181] J. Kang and S. Rhee. The Statistical models for estimating the amount of spatter in the short Circuit Transfer mode of GMAW. Welding Journal, Pages 1s-8s, January, 2001.

[182] D. V. Nishar, J. L. Schiano, W. R. Perkins and R. A. Weber. Adaptive control of temperature in arc welding. IEEE Control Systems Magazine. Volume 14(4), Pages 4-12. August, 1994.

[183] D. E. Hardt. Modeling and control of welding processes. Proc. of the Fifth Conf. on Modeling of Casting and Welding Processes. The Minerals, Metals, and Materials Society, Davos, Switzerland. Pages 287-303, 1990.

[184] N. Fourligkas and C. Doumanidis. Distributed parameter control of automated welding processes. Proceedings of 2nd IEEE Mediterranean Symposium On New Direction in Control & Automation, Pages 113-119, Maleme-Chania, Crete, Greece, June 19-22, 1994.

[185] K. Andersen, G. E. Cook, R. J. Barnett and E. H. Eassa. A class-H amplifier power source used as a high-performance welding

research tool. Proceedings of the 2nd International Conference on Trends in Welding Research, Pages 973-978. S. A. David and J. M. Vitek. Gatlinburg, TN, 1989.

[186] K. Ramaswamy, K. Andersen and G. E. Cook. New techniques for modeling and control of GTA welding. Proc. of IEEE Southeastcon 89, Pages 1250-1260. Columbia, SC. April, 1989.

[187] K. Andersen. Studies and Implementation of Stationary Models of the Gas Tungsten Arc Welding Processes. Vanderbilt University. Nashville, TN, 1992.

[188] B. F. Kuvin. Guided robots Weld army tanks. Welding Design and Fabrication, Pages 41-44. Volume 62(9), September, 1989.

[189] D. E. Hardt. Welding process modeling and re-design for control. Proceedings of 4th US/Japan Symposium on Flexible Automation, Pages 275-281 San Francisco, CA. July, 1992.

[190] G. E. Cook, K. Andersen and R. J. Barrett. Computer-based control system for GTAW. Proc. of ASME Japan/USA Symposium Flexible Automation. Pages 297-301, July, 1992.

[191] J. E. Jones. Weld parameter modeling. Proceedings of the ASM 3rd International Conference on Trends in Welding Research. S. A. David and J. M. Vitek, Gatlinburg, TN. Pages 895-898. June, 1992

[192] Y. J. Won and H. S. Cho. A fuzzy rule-based method for seeking stable arc condition under short-circuiting mode of GMA welding process. Proceeding of the Institution of Mechanical Engineers,Part1: Journal of Systems and control Engineering. Volume 206, Pages 117-125, 1992.

[193] Y. J. Won and H. S. Cho. Fuzzy predictive approach to the control of weld pool size in gas metal arc welding processes", Proceedings of ASME Winter Annual Meeting on Manufacturing Science and Engineering American Society of Mechanical Engi. ASME, New Orleans, LA. Volume 64, Pages 927-938, 1993.

[194] T. G. Lim and H. S. Cho. Estimation of weld pool sizes in GMA welding process using neural networks. Proceedings of Institution of Mechanical Engineers Part 1, Journal of Systems and Control Engineering . Volume 207(1), Pages 15-26, 1993.

[195] B. Chan, J. Pacey and M. Bibby. Modeling gas metal arc weld geometry using artificial neural network technology. Canadian Metallurgical Quarterly. Volume 38(1), Pages 43-51, 1999.

[196] Li Di, R. S. Chandel and T. Srikanthan. Static modeling of GMAW process using artificial neural networks Materials and manufacturing processes. Volume 14(1), Pages 13-35, January, 1999.

[197] A. F. Manz. The dawn of gas metal arc welding. Welding Journal, Pages 67-68. January, 1990.

[198] Anonymous. How to calculate the cost of gas metal arc welding. Welding Journal. Miami, FL. Volume 76, Pages 53-55. February, 1997.

[199] G. E. Cook. Automated arc welding-return on investment. Proceedings of Automated Arc Welding: How To Make The Right Decisions. Cleveland, OH, American Welding Society, Pages 27-45, 1980.

[200] W. R. Edwards. Controlling weld metal volume improves productivity and profit. Welding Journal, Pages 44-46. April, 1985.

[201] V. Stenke. Economic aspects of active gas metal arc welding. Svetsaren, a welding review. Volume xx(2), Pages 11–13, 1990.

[202] H. Heiro and T. H. North. The influence of welding parameters on droplet temperature during pulsed arc welding. Welding and Metal Fabrication, Pages 482-485,518. September, 1976.

[203] S. B. Jones, J. Doherty and G. R. Salter. An approach to procedure selection in arc welding. Welding Journal. Volume 56(7), Pages 19-31. July, 1977.

[204] M. J. Gellerman. How difficult is it to learn gas metal arc welding? Welding Journal, Pages 41. June, 1984.

[205] R. K. Tandon, J. Ellis, P.T. Crisp and R. S. Baker. Fume generation and melting rates of shielded metal arc welding electrodes. Welding Journal, Pages 263s-266s. August, 1984.

[206] I. Stol. Advanced gas metal arc welding process GMAW. Proceedings of the First EWI International Conference on Advanced Welding System. Edison Welding Institute. Volume 519, Pages 493–511. London, UK. P. T. Houldcroft. November, 1985.

[207] I. Stol. Development of an advanced gas metal arc welding process. Welding Journal. Volume 68(8), Pages 313s–326s. August, 1989.

[208] S. Rajasekaran, S. D. Kulkarni, U. D. Mallya and R. C. Chaturvedi. Molten droplet detachment characteristics in steady and pulsed current GMAwelding Al-Mg alloys", The Weld Institute Research Report, 1994.

[209] J. L. Pan. Study of welding arc control in China. Welding Journal, Pages 37-46", March, 1986.

[210] J.H. Chen and L. Kang. Investigation of the kinetic process of metal-oxygen reaction during shielded metal arc welding", Welding Journal, Pages 245s-251s. June, 1989.

[211] Y.S.Kim and T. W. Eagar. Temperature distribution and energy balance in the electrode during GMAW. Proceedings of the ASM 2nd International Conference on Trends in Welding Research. Gatlinburg, TN. S.A. David and J. M. Vitek, Pages 13-18. 14-18th May, 1989.

[212] B. F. Dixon. Control of magnetic permeability and solidification cracking in welded nonmagnetic steel. Welding Journal, Pages 171s-180s. May, 1989.

[213] P. A. Oberly, M. G. D'Alillio, J. E.Jones, X. Xu and D. R. White. Investigation of a blackboard artificial intelligence computer architecture for welding procedure specification and structural integrity analysis. Proceedings of the ASM 2nd International Conference on Trends in Welding Research. Gatlinburg, TN. S. A. David and J. M. Vitek, Pages 985-989, May, Pages 985-989, 1989.

REFERENCES LIST FOR CHAPTER 2

[214] S. Pandey and R. S. Parmar. Mathematical models for predicting bead geometry and shape relationships for MIG welding of aluminum alloy 5083. Proceedings of the ASM 2nd International Conference on Trends in Welding Research. Gatlinburg, TN. S. A. David and J. M. Vitek, Pages 37-41. May, 1989.

[215] C. J. Heuckroth. Wire feeders for robotic GMAW. Welding Design & Fabrication, Pages 47-49. Volume 63(9). September, 1990.

[216] N. Postacioglu, P. Kapadia and J. Dowden. Theory of the oscillations of an ellipsoidal weld pool in laser welding. Journal Physics D: Applied Physics. Volume 24, Pages 1288-1292. 1991.

[217] F. Kjeld. Gas metal arc welding for the collision repair industry. Welding Journal, Pages 39. April, 1991.

[218] T. P. Quinn and R. B. Madigan. Dynamic model of electrode extension for gas metal arc welding. Proceedings of ASM 3rd International Conference on Trends in Welding Research. Gatlinburg, TN. S. A. David and J. M. Vitek, Pages 1003-1008. June,1992.

[219] T. P. Quinn, R. B. Madigan and T. A. Siewert. An electrode extension model for gas metal arc welding. Welding Journal. Volume 73, Pages 241s-247s. October, 1994.

[220] J. P. Huissoon, D. L. Strauss, J. N. Rempel, S. Bedi and H. W. kerr. Multi-variable control of robotic gas metal arc welding. Journal of Materials Processing Technology. Volume 43(1), Pages 1-12. June, 1994.

[221] H. R. Castner. Gas metal arc welding fume generation using pulsed current. Welding Journal. Volume 74(2), Pages 59s-68s. February, 1995

[222] J. N. DuPont and A. R. Marder. Thermal efficiency of arc welding processes. Welding Journal. Miami, FL. Volume 74(12), Pages 406s-416s. December, 1995.

[223] J. Goldak, V. Breiguine and N. Dai. Computational weld mechanics: a progress report on ten grand challenges. Proceedings of the ASM 4th International Conference on Trends in Welding

Research. Gatlinburg, TN. H. B. Smartt, J. A. Johnson and S. A. David. June 5-8, Pages 5-11, 1995.

[224] H. Holm. Manufacturing state modeling and its applications. Proceedings of the ASM 4th International Conference on Trends in Welding Research. Gatlinburg, TN. H. B. Smartt, J. A. Johnson and S. A. David. June 5-8, Pages 665-675, 1995.

[225] P. G. Jonsson, J. Szekely, R. B. Madigan and T. P. Quinn. Power characteristics in GMAW: Experimental and numerical investigation. Welding Journal. Volume 74(3), Pages 93s-102s. March, 1995.

[226] P. Zhu and M. Rados and S. W. Simpson. A theoretical study of a gas metal arc welding system. Plasma Sources, Science and Technology. Volume 4(3), Pages 495-500. August, 1995.

[227] S. Adolfsson, K. Ericsson and A. Greenberg. Automatic detection of burn-through in GMA welding using a parametric model. Mechanical Systems & signal Processing Vol 10, No 5, Pages 633-651, September, 1996.

[228] H. Herold, G. Neubert, M. Zinke, U. Dilthey and A. Borner. Research into gas metal-arc welding with pulsed arc on high-alloy steels. Welding & Cutting. Volume 48(9), Pages E182-E186. September, 1996.

[229] T. Luijendijk, J. D. Zeeuw and M. P. Spikes. Calculation of the electrical resistance between contact tube and welding wire during GMA welding based on measurement of the contact force. International Journal for the Joining of Materials. Volume 8(1), Pages 1-4. March, 1996.

[230] Udo Frazz, Ronald Klier and Peter Giese. Process-integrated heat treatment as a quality assurance measure in the joining of cast iron and steel by welding. Schweissen und Schneiden / Welding and Cutting. Volume 48(4), Pages E71-E72, April, 1996.

[231] M. A. Wahab and M. J. Painter. Numerical model of gas metal arc welds using experimentally determined weld pool shapes as

the representation of the welding heat source. International Journal of Pressure vessels and Piping. Volume 73(2), Pages 153-159. September, 1997.

[232] N. Froehleke, H. Mundinger, S. Beineke, P. Wallmeier and H. Grotstollen. Resonant transition switching welding power supply. IECON Proceedings (Industrial Electronics Conference), Pages 615-620. Los Alamitos, CA, 1997.

[233] Duane Dipietro and Jim Young. Pulsed GMAW helps John Deere meet fume requirements. Welding Journal. Volume 75(10), Pages 57-58. October, 1996.

[234] K. N. Lamkalapalli, J. F. Tu, K. H. Leong and M. Gartner. Laser Weld penetration estimation using temperature measurements. Journal Of Manufacturing Science and Engineering. Volume 121(2), Pages 179-188, 1999.

[235] L. A. Jones, T. W. Eagar and J. H. Lang. Images of a steel electrode in Ar-2% Oxygen shielding during current gas metal arc welding. Welding Journal (Miami, FL). Vol 77, No 4, Pages 135s-141s, April, 1988.

[236] I. S. Kim, A. Basu and E. Siores. Mathematical models for control of weld bean penetration in the GMAW Process. Int. J. of Advanced Manufacturing Technology. Vol 12, No6, Pages 393-401, 1996.

[237] S. Rajasekaran, S. D. Kulkarni, U. D. Mallya and R. C. Chaturvedi. Droplet detachment and plate fusion characteristics in pulsed current gas metal arc welding. Welding Journal. Volume 77(6), Pages 254s-269s. June, 1998.

[238] L. A. Jones, T. W. Eagar and J. H. Lang. Magnetic forces acting on molten drops in gas metal arc welding. Journal of Physics D: Applied Physics. Volume 31(1), Pages 93-106. January, 1998.

[239] L. A. Jones, T. W. Eagar and J. H. Lang. Dynamic model of drops detaching from a gas metal arc welding electrode. Journal of Physics D: Applied Physics. Volume 31(1), Pages 107-123. January, 1998.

[240] B. J. Quimby and G. D. Ulrich. Fume formation rates in gas metal arc welding. Welding Journal (Miami, Fla). Volume 78(4), Pages 142s-149s, 1999.

[241] Y. Sugitani, Y. Kanjo and M. Ushio. High efficiency processes in automatic welding. Welding Research Abroad, Volume 45(4), Pages 32-37. March, 1999.

[242] R. Datta, D. Mukherjee, K. L. Rohira and R. Veeraraghavan. Weldability evaluation of high tensile plates using GMAW process. Journals of Materials Engineering and performance. Volume 8(4), Pages 455-462. August, 1999.

[243] Ken Michie, Stephen Blackman and T. E. B. Ogunbiyi. Twin-wire GMAW: process characteristics and applications. Welding Journal. Volume 78(5), Pages 31-34, 1999.

[244] U. Draugelates and A. Schram. Narrow-gap gas-shielded metal-arc welding of fine-grained structural steels in a vertical-up position. Schweissen und Schneiden / Welding and Cutting. Volume 51(3), Pages E42-E44, 1999.

[245] F. Tsukamoto, T. Hinata, K. Yasude and T. Onzawa. Study on welding procedure of gas metal arc welding using cored wire. Welding Research Abroad. Volume 45(11), Pages 41-49, 1999.

[246] J. Villafuerte. Understanding contact tip longevity for gas metal arc welding. Welding Journal (Miami, FL). Volume 78(12), Pages 29-35, 1999.

[247] S. Rajasekaran. Weld bead characteristics in pulsed GMA welding of Al-mg alloys. Welding Journal (Miami, Fla). Volume 78(12), Pages 397s-407s, 1999.

[248] A. Bachorski, M. J. Painter, A. J. Smailes and M. A. Wahab", Finite-Element prediction of distortion during gas metal arc welding using the shrinkage volume approach. Journal of Materials Processing Technology. Volume 92-93, Pages 405-409. August, 1999.

REFERENCES LIST FOR CHAPTER 2

[249] C. Doumanidis and Y.-M. Kwak. Geometry modeling and adaptation by infrared and laser sensing in thermal manufacturing with material deposition. American Society of Mechanical Engineers, Manufacturing Engineering Division. Volume 10, Pages 573-580, 1999.

[250] H. S. Randhawa, P. K. Ghosh and S. R. Gupta. Some basics aspects of geometrical characteristics of pulsed current vertical-up GMA weld. ISIJ International. Volume40(1), Pages 71-76, 2000.

[251] R. Brosilow. The new GMAW power supplies. Welding Design & Fabrication. Pages 22-28, Volume 60(6), June, 1987.

[252] P. Budai and B.Torstensson. A power source for advanced welding systems. Proceedings of the EWI 1st International Conference on Advanced Welding System. London, UK, Pages 424-434. P. T. Houldcroft. November, 1985.

[253] H. Yamamoto, S. Harada and T. Ueyama. Improved current control makes inverters the power sources of choice. Welding Journal. February, 1997.

[254] H. E. Essa, G. E. Cook and A. M. Wells. A high performance welding source and its application. IEEE-IAS-1983 Annual Meeting, Pages 1241-1244, 1983.

[255] T. Ogasawara, T. Maruyama, T. Saito, M. Sato and Y. Hida. A power source for gas shielded arc welding with new current weld forms. Welding Journal, Pages 57-68. March, 1987.

Chapter 3

Gas Metal Arc Welding: Sensing

Measuring, sensing, or monitoring of process parameters of a physical system is of primary importance in the application of any automatic control technique to the system. In this chapter, we discuss the sensors and sensing techniques for use in the Gas Metal Arc Welding (GMAW) process for monitoring welding parameters, welding process variables, and for calibration of welding equipment. A sensor for a welding process is an instrument or device that can be a part of the overall mechanized welding system (automatic or not) and that can transform the value of any welding variable or condition (state) of the process into a form that can be accurately measured, mostly in electrical or electronic form, such as a voltage, current, or digital signal.

3.1 Classification of Sensors

Brief accounts of the main principles and operation of sensors for welding can be found in [256, 257]. A more extensive overview of sensors for arc welding in general is found in [258]. The works [259, 260] also give a very good introduction and the state-of-the-art on sensors and their applications to arc welding processes. The development of sensors for welding processes is in general not easy due to the nature of the welding process [261]. One approach to classification of on-line sensors for welding, taken from [262], follows. An alternate classification of sensors available for arc welding is presented in [263].

1. **Before the Welding Spot:** This refers to the position and shape of the seam to be welded.

    (a) *Contact or Tactile Type:* This is perhaps one of the earliest technique for any kind of sensing of welding.

    i. Stylus or Wheel: A stylus or wheel mounting is pulled through the seam, and the position of the seam is sensed by the position of the stylus or wheel [264].

    ii. Electrical Contact with Workpiece: Another form of a tactile sensor consists of the movement of the torch towards the workpiece until it makes electrical contact with the workpiece [265]. An ultrasonic sensor having contact with the workpiece over the workpiece surface while sending the ultrasound pulses into the workpiece and measuring the reflections [266].

    (b) *Non-Contact Type*

    i. Inductive Sensors: These are used to measure the relative position of the sensor with respect to the seam [267]. Using eddy current sensors, the seam shape and position are measured for seam tracking [268].

    ii. Capacitive Sensors: These again measure the distance or position just like the inductive sensors.

    iii. Ultrasound or Ultrasonic Sensors: These sensors operate on the principle of sound measurement by indirectly measuring the time of flight for the sound to travel a distance [269].

    iv. Optical Position Sensors: The arc radiation is used as a source of light for determining the position of the seam [270].

    v. Optical Profile Sensors: These sensors use the optical principle where light, emitted by a laser to the workpiece and a sensor, is used to determine the position of the light falling on the workpiece. This was developed commercially for use in welding cars [271]. The

3.1. CLASSIFICATION OF SENSORS

same principle is extended to measure the surface profile [272, 273]

vi. Infrared Scan Sensors: Here the radiation emitted by the workpiece ahead of the weld pool is used to determine the torch position with respect to the seam [274]. Also, infrared thermography can be used for measuring weld penetration [275, 276]

2. **At the Welding Spot**

   (a) *Non-Contact*

   i. Electrical Arc Sensors): These sensors are based on of resistance of the arc and that of the electrode extension (or stick out). By moving the torch in the transverse to the direction of welding, the resistance is calculated and used to determine the center of the seam. This method is rarely used [262].

   ii. Arc Length (Optical) Sensors: Here, the basic principle is that the sensor measures the intensity of the radiation from the arc, thereby measuring the arc length [277]. In another report, the arc length is measured by deflecting the arc with an electromagnet and the arc image is captured by a charge-coupled device (CCD) camera [278].

   iii. Pool Resonance (Optical or Electrical) Sensors: We note that the weld pool oscillations are caused by current pulsing and these oscillations are different depending upon the weld penetration. With incomplete penetration, the oscillations are higher than those with full penetration (see Figure 3.1). An excellent account of weld pool oscillations, their sensing and control used in GTAW is given in [279]. The frequency of weld pool oscillations is determined by measuring the arc radiation, which changes with the frequency [280], or by measuring the arc length [281].

   iv. Pool (Spectral Lines) Sensors: The light intensity of the radiation emitted during a welding process is sensed

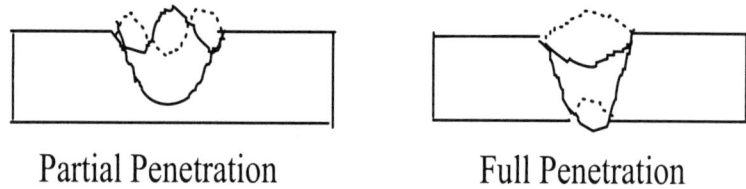

Figure 3.1: Principle of weld pool oscillations.

with an infrared detector and used for measuring the weld penetration [282].

v. **Backside Radiation (Optical) Sensors:** These sensors use the measurement of the width of the back bead in a full penetration weld [283].

vi. **Pool Geometry (Optical) Sensors:** Here the sensor uses a camera to capture the weld geometry. Pool geometry measurements are used for seam tracking [284]. Other systems used to measure the pool width are described in [285, 286, 270, 287, 288, 289]. On the other hand, [290] measures the distance from the electrode tip to front of weld pool. Other works in this category are [291, 290, 292, 293, 294].

3. **Behind the Welding Spot:** These are typically non-contact sensors used for inspection of welds already completed. A sensor is developed for inspecting workpiece profile using the principle of optical triangulation on a circle around the torch [295]. A system for three-dimensional inspection of the weld with projection of laser stripes over the weld was reported in [296].

In the remainder of this chapter we summarize various techniques available for measurement of specific weld parameters, including those listed below:

1. Conventional methods.

2. Computer-based methods.

3. Welding parameter monitoring (temperature and current).

3.2. CONVENTIONAL METHOD 99

4. Sensors for line-following and seam-tracking.

5. Arc length sensors.

6. Sensors for weld penetration control.

7. Sensors for weld pool geometry.

8. Optical sensors.

9. Miscellaneous.

3.2 Conventional Method

Basically, we have *analog* and *digital* measuring techniques. In *analog* sensors, the quantity to be measured is converted into a deflection of an indicating needle or, if the signal to be measured is rapidly changing, an oscilloscope is used. In *digital* sensing system, the analog signal is converted into digital form by using an analog-to-digital convertor (ADC). Digital instruments are more robust than analog meters.

3.3 Computer-Based Measurements

The basic principle of computer-based instrumentation is shown in Figure 3.2 [297]. The sensors from the welding process provide analog signals that are conditioned for amplification, attenuation, or filtering. The filtered analog signal is to be converted into digital from by passing it through an analog-to-digital conversion (ADC) device. Various devices such as microprocessors, programmable read-only memory (PROM), random-access memory (RAM), and input/output (I/O) devices are used in the system. The signal from the ADC is sampled at a rate higher than twice the frequency of the signal frequency in order to avoid the *aliasing* phenomena [298].

A typical welding equipment and instrumentation with a constant current power source for droplet detachment frequency measurement is given in [299].

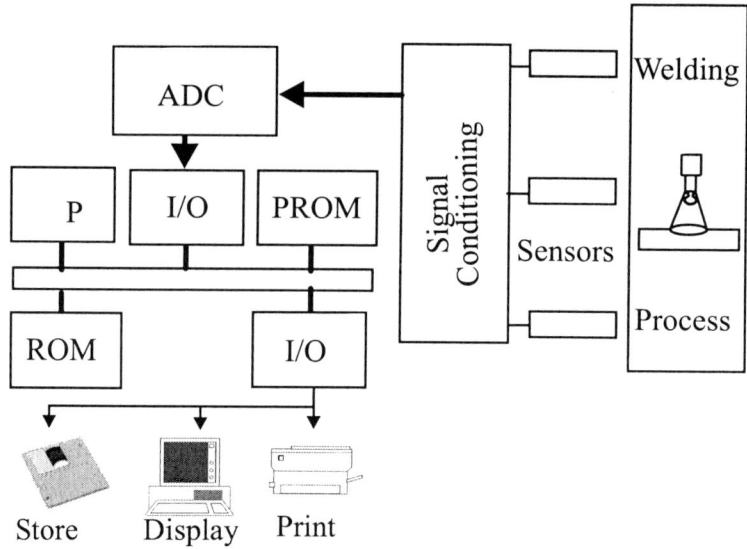

Figure 3.2: Principle of computer-based measurements.

3.4 Welding Parameters Monitoring

3.4.1 Temperature

The various types of measuring techniques or devices for measuring temperature during welding are *contact* devices such as temperature-indicating crayons or paints, bi-metal thermometers, thermocouples, and *non-contact* devices such as infrared thermometers or thermal imaging devices [300]. A very good introduction to the state-of-the-art of temperature sensors and their applications to arc welding processes are given in [259, 260].

Nishar, et al. [301] used infrared thermography for measuring weld pool temperature similar to that of [302]. The principle of infrared thermography is that all objects having a temperature above the absolute zero will radiate energy E_n, which is related to the surface temperature by Stefan's law as

$$E_n = \epsilon \sigma T^4 \qquad (3.1)$$

where ϵ is emissivity of the surface (lying between zero and unity) and

σ is the Stefan-Boltzman temperature coefficient. For temperatures resulting in arc welding, the wavelength distribution (greater than 1 μm) lies in the infrared region of the spectrum [303, 304, 305]. Somewhat inexpensive infrared sensors with shorter wavelengths (less than 1μm) have been built [306].

3.4.2 Welding Current

Sensing of welding currents which are in the range of 200 to 500 amperes, is often done by measuring the current waveform using oscilloscopes. The devices normally used for this purpose are current shunts, current transformers, or Hall effect probes.

Current shunts have low resistance values and are used to measure high currents. When connected in a circuit for measuring high welding currents, the voltage drop across the current shunt is low (50 to 200 mV), which can be easily measured by an oscilloscope. *Current transformers*, used for welding equipment using AC power supplies, are in the form of a toroid or a circular coil (sometimes with a clamp) that are placed around the conductor carrying the welding current. Here, the welding current conductor acts like the primary winding and the toroid acts like the secondary winding of a transformer. *Hall effect probes* are based on semiconductor materials that respond to the magnetic fields produced by a current-carrying conductor.

3.5 Sensors for Line Following/Seam Tracking

These sensors are meant for joint detection and/or seam tracking, which are important for robotic arc welding [307]. In particular, seam tracking has been developed for robotic arc welding processes [300, 308]. Some of the sensors used for seam tracking and other types of processes are given below.

1. Tactile sensors use a spring-loaded guide wheel that maintains a fixed relationship between the welding torch and the tracking joint.

2. Pre-weld sensing/joint location [309] may be accomplished by using the end of the GMAW electrode as a contact tube. Pre-weld

sensing with optical sensors is done for building large aluminum structures using GMAW.

3. Through-the-arc sensors [310] utilize the change in one or more electrical parameters of the arc during oscillation of the torch tip to locate the joint position. Cook [311, 312] suggested a method where the resulting welding current is proportional to electrode-workpiece distance.

4. Inductive and eddy current sensors.

5. Photoelectronic sensors [313] use a photo-emitter and photo-collector directed at a clearly defined (by using a tape if required) joint line.

6. Structured light/vision sensors [314] use a small CCD video camera to capture the image of a line of structured light projected onto the weld seam in a transverse direction. Using a strip of light, it is possible to detect a butt joint [315, 316].

7. Laser sensors are based on the laser technology.

8. Direct vision sensors [317] use a CCD video camera with the torch and may be used to view the weld area without structured light.

9. Ultrasonic sensors [318] are used to detect the infused joint line in the parent plate.

10. Chemical composition sensors use a component of the arc's spectrum to detect the presence of a particular chemical element and passing it through an on-line analysis device [319].

In the case of joint detection, a television-based optical sensing system was given in [320, 321, 317] for tracking butt joints in a GTAW process. Also see [322] for the same coaxial vision-based sensing and control in the GMAW process. Other apparatus using arc light for detecting the joint are given in [315].

3.6 Arc Length Sensors

There are basically two methods for estimating the arc length during a GMAW process. The first one is based on the measurement of signals inherent to the welding process, also called "through-the-arc" sensing. The second method is based on the use of external sensors not inherent to the process [299, 300].

3.6.1 Voltage Measurement

In general, with a constant-current power source, the arc voltage gives a good indication of arc length at a given current. In order to control the arc length, the measured arc voltage is compared with the desired (reference) voltage and the error is used to control the torch height in a typical GTAW process. A more detailed method of sensing arc length involves the process voltage [323].

3.6.2 Sound Measurement

Control of arc length may by achieved by applying oscillation frequency in the arc and measuring the sound level. The sound pressure is found to increase with arc length.

3.6.3 Laser (Range) Finders

These sensors employ a low-power solid-state laser and use triangulation techniques for sensing torch height. Laser range finder sensors can be broadly classified into those using a strip of light or a scanning beam [307]. In the first class, a laser stripe is projected onto the workpiece in front of the torch and the measurements are made by image sensors. Several works are cited under this category [324, 325, 326, 327, 328, 329, 330, 331, 332].

The second classification is the use of lasers by scanning a beam of light across the weld joint thereby obtaining the profile of the joint [272, 333, 308]. Related work can also be found in [326].

A non-contact laser profiling system has been used for measurement and prediction of weld pool shape during GMAW [334]. In this work, a detailed measurement of the full 3-dimensional weld pool shape for the GMAW process is obtained.

3.6.4 Light and Spectral Radiation Sensors

In arc light sensing, the principle used is *radiometry*, the detection and measurement of the radiant intensity of the arc [335]. Here, the incident radiant power input is converted into a proportional voltage signal [299]. The arc light, which increases linearly with arc length in GTAW, can be used to control the arc length using spectral radiation techniques [336]. An arc length sensor was developed by measuring arc current and arc light intensity using a photodiode detector [299].

A simple approach to estimate arc length using arc light for the GMAW process was patented by Johnson and Sciaky [337], where a light guide attached to the weld torch guides the light to a photodiode, the output of which is a voltage used for controlling the wire-feed speed, thus controlling the arc light intensity. In trying to investigate alternate sensors for the control of the GMAW process, such as electric field, vibration, and optical sensors, an empirical model relating arc light intensity to arc length and arc current was developed in [338]. Further, an arc length sensor, based on an arc length model incorporating simultaneous measurement of arc light intensity and arc current, was developed [299]. Here, arc length was estimated using a detector to measure arc light intensity and a current transducer to measure arc current. An expression was obtained for arc length in terms of the arc current and the detector voltage as

$$l_a = \frac{E_d - (C_0 + C_2 I^2)}{C_1 I} \quad (3.2)$$

where E_d is the detector voltage, I is the current, and C_o, C_1, and C_2 are constants dependent on various welding parameters and can be determined by a calibration procedure.

It was further found in [299] that the detector voltage is dependent on the spectral response of the detector and the orientation and location of the detector with respect to the arc source. Bonser [339] presents a multi-stripe structured light (MSSL) sensor that detects and measures the position of the saddle type weld joint formed by two small intersecting tubes.

To improve measurement accuracy, Li [340] addresses the theoretical foundation for arc light sensing. A theoretical model has been developed to correlate arc light radiation to welding parameters. It is found that the distributions of the ions of the shielding gas and the vapors of the base metal and tungsten are not even.

3.6.5 Other Works in Arc Length Sensors

A method for obtaining a signal proportional to arc length using spectral sensors based on the interference of light filters and photo triodes, is discussed in [336]. A study on the through-the-arc current and voltage signals of different types of GMAW electrodes to determine sensing strategies for the real-time control of the transfer modes is presented in [341]. This paper considers using statistical procedures, Fourier transforms, amplitude frequency histograms, peak-searching algorithms, and smoothing techniques.

An arc hydrogen sensor (AHS) is developed, built, and demonstrated in [342] in a laboratory environment. It shows that the hydrogen levels down to 1000 ppm, loss of shielding gas, and the presence of grease on the part are detectable.

A simple through-the-arc sensing technique for real-time monitoring of weld quality is proposed in [343, 344], where current and voltage records for pulsed GMAW were captured and correlated with high speed images of the arc to obtain any defective changes in loss of shielding gas, contact-tip-to-work distance wear. Dynamic analysis of arc length and the development of a sensor for measurement of high-frequency weaving is given in [345].

3.7 Sensors for Weld Penetration Control

These techniques are developed mainly for situations where a full penetration. In [266], the geometry of the weld pool during the GMAW process was detected by using high-frequency sound waves. Further, an expert system was used for automatic identification of molten/solid interface geometries. Also see other works by the same group [346, 347, 348, 349].

An infrared camera is used to obtain temperature profiles for the GTAW process and to determine the relationship between weld penetration variations and plate temperature distributions in [350]. The work in [351] describes a proportional-integral-derivative-based control system for controlling weld penetration based on coaxial viewing by indirectly controlling the weld pool width. The development of a laser fiber bundle array and electromagnetic acoustic transducer (EMAT) for the generation and reception of ultrasound for on-line weld penetration measurement is discussed in [352]

3.7.1 Back-Face Sensing

Back-face sensing involves sensing from the underside (below or back) of the workpiece. The energy of certain wavelengths of radiation emitted from the back of the weld pool is a function of the weld bead area or width [353]. The signal from the sensor is used to control the process variables such as pulse duration in a pulsed GTAW process. This technique was also used with pulsed metal transfer in GMAW process to control bead width [354].

Radiation Sensing

Here the radiation from the back surface of the weld pool is collected by an optical sensor in the form of an optical fiber connected to a camera or an optical pyrometer camera. The radiation frequency is directly proportional to temperature [353, 355, 356].

Shadow Motion Sensing

This technique is based on the sagging effect on weld pool, which occurs at full penetration. The sagging is detected from the underside of the work piece, as shown in Figure 3.3. A laser beam is passed along the back surface of the workpiece. The beam is partially blocked or shadowed by the sagging weld pool, with the sagging proportional to the weld current [279].

3.7.2 Front- or Top-Face Sensing

Front face sensing involves sensing from the top (torch) side. This can be done using light, sound, infrared, radiological, or other energy sources. Penetration monitoring from the top face (front face) of the weld pool involves the application of a short-duration, high-current pulse, which excites the pool and produces oscillation [280].

3.7. SENSORS FOR WELD PENETRATION CONTROL

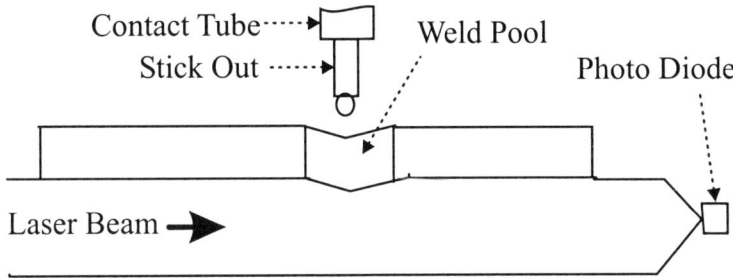

Figure 3.3: Schematic of laser shadow motion sensing method.

Radiographic Sensing

In radiographic sensing, X-rays coming on to the top side of the workpiece are reflected two times at the bottom of the weld pool and the resulting difference in intensity is passed through a digital image processor and a computer for automatic control of weld penetration. Because the X-ray source is located at the top of the work piece, it is sometimes called top-face sensing [357, 358].

Ultrasonic Sensing

In a simple configuration, two transducers, one acting as transmitter T and the other as receiver R, are placed on both sides of the welding head, as shown in Figure 3.4[279]. A linear encoder, mounted on the

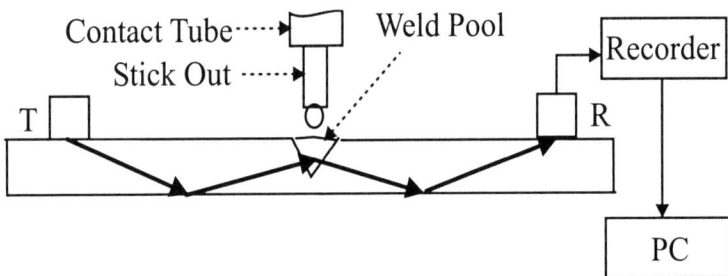

Figure 3.4: Simplified schematic of ultrasonic sensing method

welding head, receives the ultrasonic signals which are passed on to a transient recorder and a personal computer (PC) for analysis, display, and automatic control [359, 266]. Alternatively, the signals can be collected by a video system. Additional works were reported with regard to contactless transducers and laser sound generation [360, 361].

Ultrasonic methods using a dual-element piezoelectric transducer were used to detect discontinuities in the weld pool in a GMAW process [361]. Also, expert system principles were used to design intelligent controller for manipulating the welding parameters. A method consisting of a pulsed laser for ultrasound generation and an electromagnetic acoustic transducer (EMAT) for ultrasound reception was developed by Carlson et al [362] for detection of weld defects. Other works by the same group can be found in [346, 347, 266, 348, 349].

Other works on ultrasonic sensing include a technique developed in [363] using ultrasonic pulse-echo measurements to determine the weld pool dimensions. Another sensor capable of simultaneously controlling weld joint penetration and joint tracking is developed in [359]. A very good introduction to the state-of-the-art of ultrasonic sensors and their applications to arc welding processes can be found in [259, 260, 364]. A novel seam tracking technology based on high-frequency ultrasound is developed in [365] to achieve high accuracy in weld seam identification.

Infrared Sensing

For sensing the surface temperature of the weld pool, an infrared camera is mounted towards the weld bead. Perturbations in weld penetration are detected from the measured temperature gradient [366, 275, 367]. A top-side 3-D infrared sensing technique to be used in closed-loop control of weld full penetration in the GTAW was presented in [368]. Using front-end (or side) infrared sensors, an on-line scheme for monitoring weld geometry was presented for both GTAW and GMAW processes by Banerjee, et al. [369].

The application of infrared sensing and computer image processing techniques for dynamic control of joint penetration parameters is presented in [275]. A linear relationship between the weld bead width and the infrared thermal image profile was established. Weld faults such as arc misalignment, variation in penetration, and impurities are found to produce distinct thermal distributions of the weld pool surface temperature, as measured by scanning infrared cameras in [274].

3.7. SENSORS FOR WELD PENETRATION CONTROL

One of the earliest works investigating the feasibility of using infrared sensing devices to determine the heat flow pattern with changes in welding parameters for both GTAW and GMAW processes is given in [370]. Other works include an ultrasonic sensor used for on-line weld penetration measurement in [371].

A feasibility study has been done in [276] to determine if infrared thermography could be used to detect perturbations in the arc welding process that result in defects such as arc misalignment, puddle impurities etc. An investigation on the feasibility of using infrared sensor (camera) to monitor the molten weld pool during GTAW and GMAW processes is discussed in [372]. Further investigations on the use of infrared thermography to sense the position of the arc and the penetration depth of the weld can be found in [373, 367].

In [374], an infrared camera was used to record temperature gradients surrounding the welding torch and transmit the images to a central computer for image processing . An infrared camera has also been used to obtain temperature profiles for GTAW process in [350]. Infrared sensing has been used for thermal distribution changes associated with GTAW to identify and correct for weld joint offsets for butt joints in [375].

A very good introduction to the state-of-the-art on infrared sensors and their applications to arc welding processes can be found in [259, 260]. Related works on the topic of infrared sensing can also be found in [373, 275, 288].

Trace Element Method

In this method, prior to the start of the welding process, tracers are attached to the back face of the weld piece. When full penetration reaches the traces, they are melted and mixed with the weld pool and transferred by convection to the surface of the weld pool, where they are detected by a spectral emission collector and processed by a computer for data analysis. With partial penetration, the tracers are not melted and hence are not detected. The choice of the tracer is an important factor in this technique [279].

Weld Pool Depression Method

In this method [378], the weld pool surface sag is used to estimate the weld penetration. A sagging weld pool indicates a full penetration whereas no sagging shows partial penetration [376, 377].

3.7.3 Weld Pool Oscillation

Weld pool oscillations are caused by high frequency external forces on the weld pool. It was first suggested that the ripple formation in solidified welds is explained by the oscillatory behavior of the weld pool [379]. Droplet transfer can also generate weld pool oscillation during GMAW [300]. It is worth noting that the weld pool oscillation frequency will be influenced by the droplet frequency [279]. The measurement of forced oscillation of the weld pool via the arc voltage is possible but noise seems to be a problem [380].

Weld pool oscillations can also be induced by current pulsing and monitored using optical sensing. This approach is applied for the GTAW process in [381]. In particular, the oscillations are induced by a phase-locked loop (PLL) which consists of a phase detector, low-pass filter, and oscillator [382, 383, 384, 385]. A simplified block diagram of the PLL pulsing system is shown in Figure 3.5 [].

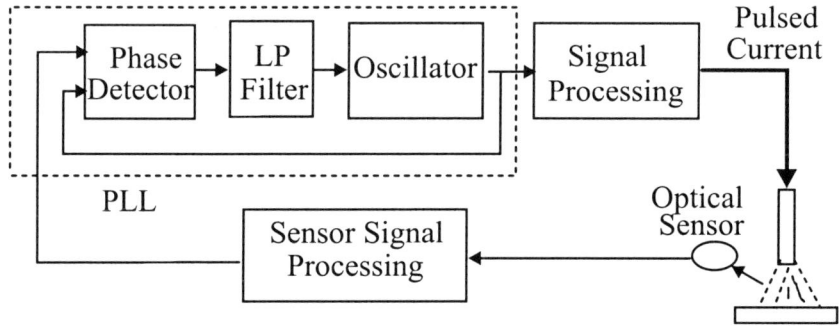

Figure 3.5: Simplified schematic of phase locked loop method for pulsing

Weld pool oscillations were also studied in [386]. For further work on weld pool frequency and its monitoring, see [387] for a weld penetration control and monitoring system, called MELODY, that is based

3.8. SENSORS FOR WELD POOL GEOMETRY

on weld pool frequency and uses a fiber optic sensor. Also, see [388] for a discussion on the role of weld pool oscillations during arc welding, with particular reference to the types of oscillation modes, oscillation frequency, weld pool geometry, etc. Further, weld pool oscillations were used for weld penetration sensing and control, in particular for GTAW [279].

3.7.4 Droplet Transfer Frequency

There are several droplet transfer and frequency sensing techniques that have been developed over the years for GMAW, primarily focused on measurement of arc current and/or voltage [341, 389], or air-borne acoustic sensing [390, 391, 392]. These methods were developed for real-time implementation in short-circuit and globular metal transfer modes. Alternatively, it was shown that the use of static transfer mode maps could be used to *off-line* by using average values of process voltage and current from post-weld inspection of high-speed videos [393].

Droplet transfer frequency was obtained from variations in arc length, measured by an arc light sensor (detector), caused by droplet detachment [338, 299]. The sensor is a commercially-available photodiode [394]. It was found that the droplet frequency was a function of current, wire feed speed, and electrode extension.

It was observed that in the case of spray transfer mode, the droplet frequency is directly proportional to current [341, 395] and that the globular and short-circuit transfer modes can be controlled by current and voltage sensing [341, 389, 390]. Also, see [396] for sensing of droplet transfer using an ER100S-1 electrode, and [338] for a study on the use of arc light intensity sensors in conjunction with voltage and current sensors to determine the droplet detachment.

3.8 Sensors for Weld Pool Geometry

The detection of weld pool geometry can be done using optical sensing techniques such as photodiodes or charge-coupled devices (CCD) [397, 398]. It is found that only 4% of the energy can be detected. In order to overcome this, one can momentarily extinguish the arc to record the CCD array response [290]. Alternatively, a narrow band filter can be used [399, 331]. Also, one can mount the vision apparatus such that the optical axis is aligned with the electrode [286, 321]. In measuring the weld pool, which partly includes the arc, there is a large difference in emissivity between the liquid and solid metal [307].

A measurement system was developed for weld pool geometry in GMAW process by observing radiation emitted by the weld pool using a solid state imaging camera based on CCD [262]. A dedicated computer is used to perform the image analysis based on edge detection rather than thresholding the image. Advanced imaging techniques are used for on-line measurement of weld pool geometry in [400, 401, 284, 287, 321].

Another class of sensors for measuring weld pool geometry is infrared (IR) vision sensors which detect the changes in radiation on the surface of the weld pool with or without correlating them to pool temperature [370]. Further applications of IR sensors for extracting geometrical features of weld pool are found in [402, 288].

Using thermographic images of the weld pool, impurities and variations in joint geometry were detected in [274]. Sometimes, a complete knowledge of temperature distribution is not necessary, but the temperature at only two points at equal distance from the torch may be sufficient for control purposes [403].

A vision-processing algorithm was developed to compute weld pool (puddle) geometry parameters from the noisy image of the molten pool [404]. Infrared thermography was used in determining the puddle geometry and cooling rates for the GMAW process in [305, 303].

A method for determining the optimum sensor location to measure weldment surface temperature, which has a close relation with weld pool size in the GMAW process was given in [405]. It was found that the correlation significantly changes with measurement location and the optimal location occurs at the maximum correlation value.

Kovacevic and Zhang have developed several novel sensing methods for weld pool sensing. In particular, in [406] the specular reflection of pulsed laser stripes from the mirror-like weld pool surface was captured by a CCD camera and, for monitoring and control of the process, the captured image was analyzed to identify the torch and electrode. Finally, the reflection image pattern and the pool boundary were used to control the weld penetration [407]. Extensive results on this and related topics by these researchers are found in [408, 409, 410, 411, 412,

3.9. OPTICAL SENSORS 113

413, 414, 412, 415, 416, 417, 418, 419, 420, 421, 422, 423, 424, 425, 426, 427, 428]

Other works on this topic are listed below. Two techniques of measuring weld pool surface size during welding are introduced in [397, 398]. One is based on self-scanning photo-diode arrays with momentary arc extinction suitable for weld pool control. The other uses TV cameras and arc filtering suitable for weld pool monitoring.

A method of computer analysis for two-dimensional imaging of GTAW pools as acquired by a coaxial viewing system was developed in [429]. Real-time radiography (x-ray and image intensifier units) was developed in [357, 358, 294] for observation of weld pool volume and the heat-affected zone during the welding process.

Investigations to determine the feasibility of using ultrasonic sensing technique for weld pool depth and penetration are reported in [366, 318]. A study on the reliability of vision sensors for recording the variation in arc noise by considering the reflection of the base metal surface, which is modeled using a bidirectional reflectance-distribution function (BRDF), is presented in [430]. Also, using a thermographic imaging technique, it may be possible to assess the temperature profile of a weld joint thereby controlling the metallurgical properties of the weld [402].

3.9 Optical Sensors

A temperature control system was designed for a consumable electrode GMAW process using an inexpensive optical system for measuring weld temperatures [301]. Development of a control system using an optical sensor (photodiode array camera) for measurements of the groove geometry in front of the welding head is given in [431].

The use of coaxial viewing for adaptive welding control of the GMAW process is demonstrated in [432]. Coaxial viewing incorporates a camera on the torch such that the torch axis is aligned with the optical axis. An integrated optical sensor (IOS) for feedback control in the GMAW process is developed in [433]. The IOS consists of three readily available components: a charge-coupled device (CCD) camera, a diode laser, and a processing computer and measured in almost real time the parameters such as weld pool position and width, and weld bead centerline cooling rate. In [434], the surface tension of detached liquid drops in pulsed GMAW was determined from the period of the oscillations initiated by the detachment event.

Lee, in [435], discusses vision sensors using optical triangulation, which have been widely used for automatic welding systems in various ways. In this study, the reliability of vision sensors was analyzed for the variation of the arc noise by considering the reflectance of the base metal surface. The property of the surface reflection of the base metal was modeled using the bi-directional reflectance-distribution function (BRDF).

3.10 Sensors for Quality Control

There are some metallurgical properties, such as porosity and cracking that need to be monitored and controlled on-line to ensure proper weld quality. Some of the visual indications characterizing the weld quality are cracks, porosity, undercuts, micro fissures, etc. [436]. An automated system has been developed for visual inspection of weld beads and evaluation of their quality [437, 438], using the human techniques of observation, inspection and evaluation. A discussion on the stochastic behavior of arc welding signals with respect to monitoring for quality assurance requirements can be found in [439], where a portable arc data analyzer was described.

Statistical process control (SPC) is defined as *the monitoring and analysis of process conditions using statistical techniques to accurately determine process performance and prescribe preventive or corrective actions as required* [440]. In many cases, if the welding process is operated under normal conditions and sufficient data is collected, the evaluated values fall within a standard probability distribution. Dynamic resistance monitors are built incorporating statistical process control techniques.

3.11 Intelligent Sensing

In this section we assemble works on sensing techniques using artificial neural networks (ANN), fuzzy logic (FL), knowledge-base systems (KBS), and expert systems (ES) [392]. In [266], an expert system was

3.12. OTHER ISSUES ON SENSING

used for automatic identification of molten/solid interface geometries during ultrasonic sensing of the GMAW process. Also, see other works by the same group [347, 348, 349, 361] as well as [441] for the development of an expert system to diagnose discontinuities in GMAW. Also, see [360] for the application of the expert-system approach to analyzing ultrasonic measurements of the weld pool geometry.

An ANN-based technique was developed in [392] for classifying the acoustic signals of acceptable and unacceptable welds obtained by the GMAW process. Kovacevic and Zhang [426] proposed a neuro-fuzzy model for weld fusion to infer the back-side bead width from the pool geometry, using the knowledge of the skilled worker (formulated into a set of fuzzy rules) and a neural network for learning to adjust the parameters in the fuzzy model [442]). In the experimental setup, the weld pool image is captured by a CCD camera and processed through an image processing unit. Then a neurofuzzy estimator provides the weld bead geometry, which are incorporated into a feedback algorithm to achieve the *desired* bead geometry. For extensive results on this and related topics by this group, see [408, 411, 443, 444, 416, 407, 428, 427].

Other works on this topic are given below. The development of a neural network-based system to perform automated visual seam tracking in arc welding is proposed in [445]. The use of neural networks with infrared sensing of weld penetration control is reported in [446]. An acoustic sensing method was developed using neural networks in [392].

The application of a pattern recognition technique to interpret the arc weld images for coaxial weld vision based process control is presented in [447]. The application of neural networks for identification of weld defects is discussed in [448]. Du, et al. in a two part series [449, 450], presented a systematic study of various monitoring methods suitable for automated monitoring of manufacturing processes and their applications to some specific processes, including arc welding. using techniques of pattern recognition, neural networks, fuzzy systems, decision trees, and expert systems.

3.12 Other Issues on Sensing

Further, we review additional works on sensing for the GMAW process. In [451], a study was made about the presence of strong environmental noise from arc light and molten-particle emission, leading to the choice of effective noise-reduction techniques. In a keynote address, Araya and Saikawa [452], gave an excellent account of activities (particularly in Japan) in sensing and adaptive control of arc welding.

Also, see some of the earlier works by Arata, et al. [401] for the application of a digital picture processing technique used in automatic control of arc welding processes. In [453], the wide use of the arc sensor to detect the weld seam by monitoring welding current or voltage variation during weaving in GMAW is discussed. In this work, the arc light intensity and welding resistance are compared when Argon and Carbon-Dioxide gas are used for shielding. A method for detecting flaws in automatic, constant-voltage gas metal arc welding using the process current and voltage signals was developed in [454]. Seven algorithms process the current and voltage signals to give quality parameters.

It is possible to have a combination of the above sensing techniques for obtaining a better indication of the process and its variables. For example, combining the torch displacement sensor with through-the-arc measurements of voltage and current may be used to distinguish between the wire feed slip and torch height variation [300].

Spectrographic techniques were used to detect changes in chemical composition of the arc in [455]. A discussion on the need for special sensors for robotic arc welding was made in [456]. The Univision II sensing system, which is basically an optical seam tracking system consisting of a sensor, a TV camera, and an image processor, was described in [351].

An excellent tutorial and survey type of discussion on modeling, sensing, and control of welding processes can be found in [457]. The importance of sensing for automated welding is discussed in [458]. The work on the study of arc sensors for GMAW is presented in [459, 460]. In this work, the welding current signal was fit to a curve that is inversely proportional to the trace of the contact tip-to-workpiece distance by using a quadratic curve fitting method in order to extract useful information on the welding gun position from the welding current signal.

A direct-view vision system was developed in [461] to track joints in both GTAW and GMAW processes, based on the analysis and un-

3.12. OTHER ISSUES ON SENSING

derstanding of the radiation energy from the weld region. Sensors and control systems in arc welding used/developed mostly in Japan are presented in an edited Book [462]. An excellent introduction to sensing, its classification and its applications to arc welding processes are presented in [259].

A state-of-the-art review of sensors and their applications to arc welding processes are given in [260]. An interesting discussion of results obtained by a questionnaire on sensor applications in welding processes is given in [463].

Other related works can be found in the section on sensing, control and automation in [464]. Future trends for welding sensors are discussed in [465]. Chen, et al., [466] used an industrial TV camera as a sensor to measure the weld face width of the weld pool, by employing computer imaging techniques. In another work, on-line, computer-based monitors provide quality assurance, which enhances the reliability of the GMAW process and reduces the need for post-weld testing [467].

3.13 Classification of References by Section

Here, we provide a table containing the various references according to each section of this chapter. This will provide a ready reference to the interested reader to search for relevant references in each section.

3.13. CLASSIFICATION OF REFERENCES BY SECTION

Table 3.1: Section by Section List of References

Section	Reference Numbers
3.1 Classification of Sensors	[256]-[273]
3.2 Conventional Method	
3.3 Computer-Based Measurements	[297]-[299]
3.4 Welding Parameters Monitoring	[295],[296],[300]
3.5 Sensors for Line Following/Seam Tracking	[300]-[316]
3.6 Arc Length Sensors	[269],[299]-[302], [317]-[339]
3.7 Sensors for Weld Penetration Control	[263],[271]-[273], [276],[277],[285], [295],[296], [299], [300],[318],[337], [340]-[363], [369]-[396]
3.8 Sensors for Weld Pool Geometry	[260],[271],[281], [283]-[285],[287], [291],[301], [312], [315],[331],[351], [352],[360],[364], [366],[370], [397]-[430]
3.9 Optical Sensors	[362],[431]-[435]
3.10 Sensors for Quality Control	[340]-[436]
3.11 Intelligent Sensing	[263],[341]-[343], [354],[355],[391], [407],[408], [411], [416],[426]-[428], [441]-[450]
3.12 Other Issues on Sensing	[295],[296],[300], [345],[401], [451]-[467]

References List for Chapter 3

[256] R. B. Madigan. Ways to keep torches in seams. Welding Design and Fabrication. Volume 60(10), Pages 48-50, 1987.

[257] U. A. Dilthey. Sensor technique for welding robots: some of the developments and trends. Intl. Inst. of Welding. XI-1129-89, 1989.

[258] A. Matsunawa et al. Sensors and Control Systems for Arc Welding. Japan Welding Society. Tokyo, Japan, Volume IIW Doc. XII-1220-91, 1991.

[259] H. Nomura. Introduction to sensing systems and their application in arc welding processes. Sensors and Control Systems in Arc Welding. Chapman & Hall. London, UK. English Translation of the Original 1991 Japanese Edition. Chp 1, Pages 1-17, 1994.

[260] H. Nomura. State-of-the-art review of sensors and their application in arc welding processes. Sensors and Control Systems in Arc Welding. Chapman & Hall. London, UK. English Translation of the Original 1991 Japanese Edition. Chapter 2, Pages 18-43, 1994.

[261] S. B. Jones and G. Starke. Applications, advantages and approaches for image analysis in the control of arc welding. Esprit '84: Status Report of Ongoing Work. J. Roukens and J. F. Renuart. Elsveir Science Publishers. Amsterdam, The Netherlands, Pages 425-455, 1985.

[262] H. C. Wezenbeek. A System for Measurement and Control of Weld Pool Geometry in Automatic Arc Welding. Technische Univ.

Eindhoven Netherlands, Dept. of Electrical Engineering. Eindhoven, The Netherlands, 1992.

[263] H. B. Smartt. Intelligent sensing and control of arc welding. Proceedings of the ASM 3rd International Conference on Trends in Welding Research. Gatlinburg, TN. June, S. A. David and J. M. Vitek. Pages 843-851, Keynote Address. June, 1992.

[264] W. Faber and D. Lindenau. Taktile sensoren zur automatisierung in der schweib-technik. ZIS-Mitteilungen, Volume 27(12), Pages 1290-1296, 1985.

[265] C. P. Cullen. An adaptive robotic welding system using weldwire touch sensing. Welding Journal, Volume 67(11), Pages 17-21, 1988.

[266] N. M. Carlson and J. A. Johnson. Ultrasonic sensing of weld pool penetration. Welding Journal, Volume 67(11), Pages 293s-246s. November, 1988.

[267] F. Goldberg. Inductive seam-tracking improves mechanized and robotic welding. Proceedings of the Intl. Conf. on Automation and Robotization in Welding and Allied Processes. Intl. Inst. of Welding and Pergamon. Oxford, UK, 1985.

[268] D. Placko, H. Clergeot and F. Monteil. Seam tracking using a linear array of eddy current sensors. Proceedings of the 5th Intl. Conf. on Robotic Vision and Sensory Controls. N. J. Zimmerman. Amsterdam, The Netherlands. Pages 557-568, 1985.

[269] B. Maqueira, C. Umeagukwu and J. Jarzynski. Robotic seam tracking of weld joints through the use of ultrasonic sensors. Proceedings of Intl. Workshop on Industrial Applications of Machine Vision and Machine Intelligence. Tokyo, Japan, Pages 291-296, 1987.

[270] J. C. Bonvalet, Y. Launay and C. Philip. Adaptive welding control using a CCD sensor. Proc. of Intl. Conf. on Automation and Robotization in Welding and Allied Processes. Intl. Inst. of Welding and Pergamon. Oxford, UK, Pages 365-367, 1985.

REFERENCES LIST FOR CHAPTER 3

[271] T. Sthen and T. Porsander. An adaptive torch position system for welding of car bodies. Proc. of the 3rd Intl. Conf. on Robot Vision and Sensory Controls. Cambridge, MA, Pages 607-613, 1983.

[272] G. L. Oomen and W. J. P. A. Verbeek. A real-time optical profile sensor for robot arc welding. Proceedings of Third International Conference on Robot Vision and Sensory Controls RoViseC3. Cambridge, MA. Volume 449, Pages 67-71. November, 1983.

[273] D. C. Verdon, D. Langley and M. H. Moscardi. Adaptive welding control using video signal processing. Proceedings of the International Conference on Developments in Mechanized, Automated and Robot Welding, Pages P291. London, UK. November, 1980.

[274] B. A. Chin, N. H. Madsen and J. S. Goodling. Infrared thermography for sensing the arc welding process. Welding Journal, Volume 62(9), Pages 227s-234s. September, 1983.

[275] W. H. Chen a nd B. A. Chin. Monitoring joint penetration using infrared sensing techniques Welding Journal. Volume 69(4), Pages 181s-185s, 1990.

[276] M. A. Khan, N. H. Madsen, J. S. Goodling and B. A. Chin. Infrared thermography as a control for the welding process. Optical Engineering. Volume 25(6), Pages 799-805, 1986.

[277] R. T. Deam and P. N. Drew. Relationship between arc light, current and arc length in TIG welding. Proc. of Intl. Conf. on Advances in Joining and Cutting Processes. Harrogate, UK, 1990.

[278] X. Q. Chen and J. Lucas. A fast system for control of narrow gap TIG welding. Proc. of Intl. Conf. on Advances in Joining and Cutting Processes. Harrogate, UK, 1990.

[279] A. J. Aendenroomer. Weld Pool Oscillation for Penetration Sensing and Control. Technical School of Delft University. Delft, The Netherlands, 1996.

[280] R. J. Salter and R. T. Deam. A practical front face penetration control system for TIG welding. Proceedings of the Conference on Developments in Automated and Robotic Welding. The Welding Institute, London, UK, 1986.

[281] Y. H. Xiao and G. den Quden. A study of GTA weld pool oscillation. Welding Journal, Volume 69(8), Pages 289s-293s, 1990.

[282] D. A. Dornfeld, M. Tomizuka and G. Langeri. Modeling and adaptive control of arc welding processes. Measurement, Control in Batch Manufacturing, ASME. D. E. Hardt. New York, NY. Pages 53-64, November, 1982.

[283] A. Suzuki and D. E. Hardt. Application of adaptive control theory to in-process weld geometry regulation. Proceedings of the 1987 American Control Conference. Pages 723-728, 1987.

[284] K. Inoue. Image processing for on-line detection of welding process (Report 11) - binary processing for the image of arc welding process. Transaction JWRI. Volume 9(1), Pages 27-30, 1980.

[285] A. R. Vorman and H. Brandt. Feedback control of GTA welding using puddle width measurement. Welding Journal. Volume 55(9), Pages 742-749, 1976.

[286] R. A. Richardson, D. A. Gutow and S. H. Rao. A vision based system for arc weld pool size control. Measurement and Control for Batch Manufacturing. D. E. Hardt, ASME. New York, NY. Pages 65-75, November, 1982.

[287] K. Inoue. Simple binary image processor and its application to automatic welding. Proc. of Joint Automatic Control Conference. 1080.

[288] G. Begin and J. P. Boillot. Welding adaptive functions performed through infrared(IR) simplified vision schemes. Proceedings of the 3rd International conference on Robot Vision and Sensory Controls RoViSeC3. Cambridge, MA. Volume 449, Pages 328-337. November, 1983.

REFERENCES LIST FOR CHAPTER 3

[289] M. Yamamoto, Y. Kaneko, K. Fujii, T. Kumazawa, K. Ohishima, G. Alzamora, T. Kubota, F. Ozaki and S. Anzai. Adaptive control of pulsed MIG welding using image processing system. Rec. of Conf. of the 23rd IEEE Industry Applications Society. Pittsburgh, PA, Pages 1381-1386, 1988.

[290] F. Bruemmer and R. Niepold. Understanding and controlling arc welding processes by monitoring the melting pool with an optical sensor. Proc. of the Intl. Workshop on Industrial Applications of Machine Vision and Machine Intelligence. Tokyo, Japan. Pages 285-290, 1987.

[291] R. Niepold and F. Brümmer. A visual sensor for seam tracking and on-line process parameter control in arc-welding applications. Proc. of 14th Intl. Symp. on Industrial Robots and 7th Intl. Conf. on Industrial Robot Technology, Pages 375-385, 1984.

[292] F. Nadeau, J. Blain and M. Dufour. Computerized system automates GMA pipe welding. Welding Journal. Volume 69, Pages 375-385, 1990.

[293] L. M. Sweet. Sensor-based control systems for arc welding robots. Robotics and Computer-Integrated Manufacturing. Volume 2(2), Pages 125-133, 1985.

[294] S. I. Rokhlin and A. C. Guu. Computerized radiographic sensing and control of an arc welding pro-cess. Welding Journal, Pages 83s-97s. March, 1990.

[295] D. N. Karastojanov and G. N. Nachev. Adaptive control of industrial robots for arc welding. Proc. of the 9th Triennial World Congress of the IFAC, Pages 2405-2410. Budapest, Hungary. July, 1985.

[296] J. E. Agapakis, N. Wittles and K. Masubuchi. Automated visual weld inspection for robotic welding fabrication. Proc. of the Intl. Conf. on Automation and Robotization in Welding and Allied Processes. September, Pages 151-160, 1985.

[297] J. Norrish. Computer based instrumentation for welding. Proceedings of Conference on Computer Technology in Welding. The Welding Institute, Cambridge, UK. June, 1988.

[298] C. L. Phillips and H. T. Nagle. Digital Control System Analysis and Design, Third Edition. Prentice Hall. Englewood Cliffs, NJ, 1995.

[299] R. B. Madigan. Control of gas metal arc welding using arc light sensing. Colorado School of Mines. Golden, CO, 1994.

[300] J. Norrish. Advanced welding processes. Institute of Physics Publishing. Bristol, UK, 1992.

[301] D. V. Nishar, J. L. Schiano, W. R. Perkins and R. A. Weber. Adaptive control of temperature in arc welding. IEEE Control Systems Magazine. Volume 14(4), Pages 4-12, August, 1994.

[302] W. E. Lukens and R. A. Morris. Infrared temperature sensing of cooling rates for arc welding control. Welding Journal. Volume 61(1), Pages 27-33, 1982.

[303] J. L. Schiano. Feedback control of two physical processes: design and experiments. University of Illinois. Urbana, IL. DC-148, UILU-ENG-93-2217, 1993.

[304] B. W. Greene. Arc current control of a robotic welding system: modeling and control system design. Illinois Univ. at Urbana-Champaign, Coordinated science Lab.. Champaign, IL. DC-114-UILU-ENG-89-2227, 1990.

[305] J. L.Schiano, D. E. Henderson, J. H. Ross and R. A.Weber. Image analysis of puddle geometry and cooling rate for gas metal arc welding control. Proc. of the Inst. of Soc. Opt. Eng., Midwest Tech. Conf. Chicago, IL, 1991.

[306] R. E. Sampson, G. Suits and C. B. Arnold. Analysis of sensors for application to welding control. University of Michigan. Ann Arbor, MI. N00167-84-K-0036, 1985.

[307] P. Sicard and M.D. Levine. An approach to an expert robot welding system. IEEE Transactions on Systems, Man and Cybernetics. Volume18(2), Pages 204-222, March/April, 1988.

[308] N. Nayak and A. Ray. Intelligent Seam Tracking for Robotic Welding. Springer-Verlag. London, UK. Advances in Industrial Control, 1993.

[309] M. L. Kohn, S. A. Ramsay, D. T. Damon and B. W. Folkening. A robotic cell for welding large aluminum structures using a two-pass optical system. First International Conference on Advanced Welding systems. Houldcroft,P.T. Welding Inst, Abingdon, Cambridge, UK. Pages 17-29, 1985.

[310] R. Hughes. Arc guided robot plasma arc welding. Proceedings of the 1st International Conference on Advanced Welding Systems. The Welding Institute, London, UK. November, 1985.

[311] G. E. Cook. Through-the-arc sensing for arc welding. Proceedings of the 10th NSF Conference on Production Research and Technology. Pages 141-151, February28-March2, 1983.

[312] G.E. Cook. Robotic arc welding: Research in sensory feedback control. IEEE Transactions on Industrial Electronics. VolumeIE-30(3), Pages 252-268, August, 1983.

[313] C. Tan and J. Lucas. Low cost sensors for seam tracking in arc welding. Proceedings of the 1st International Conference on Computer Technology in Welding. The Welding Institute, London, UK. June, 1986.

[314] J. E. Agapakisetal. Joint tracking and adaptive robotic welding using vision sensing of the weld joint geometry. Welding Journal. November, 1986.

[315] Y. Arata and K. Inoue. Automated control of arc welding (report II)-optical sensing of joint configuration. Transaction. Jap. Welding Res. Institute (JWRI). Volume2(1), Pages 87s-101s, 1973.

[316] K. Inoue, H. Akashi, M. Tamaoki, Y. Shibataand, and H. Arata. Automatic control of arc welding (report 11). Transaction Japanese Welding Research Institute(JWRI). Vol 9, No 1, Pages 31-37, 1980.

[317] R. W. Richardson, A. Gutow, R. A. Anderson and D. F. Farson. Coaxial arc weld pool viewing for process monitoring and control. Welding Journal. Pages 43-50, March, 1984.

[318] R. R. Stroud and T. J. Harris. Seam tracking butt and filler welds using ultrasound. Joining of Materials. Volume 2, 1990.

[319] Anonymous. Composition Controlled Sensing Technology. U-Weld Automation. Rockville, MD, 1986.

[320] R. W. Richardson. Review of the state-of-the-art of adaptive control for the gas tungsten and plasma arc welding processes. Air Force Materials Laboratory, Wright-Patterson Air Force Base. Dayton, OH, 1981.

[321] R. W. Richardson and R. A. Anderson. Weld butt joint tracking with a coaxial viewer based weld vision system. Proceedings of ASME Winer Annual Meeting on Control of Manufacturing Process and Robotic Systems. ASME, D. E. Hardt and W. J. Book. Boston, MA. Pages 107-119, November, 1983.

[322] R. W. Richardson and C. C. Conrardy. Coaxial vision-based control of GMAW. Proceedings of ASM 3rd International Conference on Trends in Welding Research. Gatlinburg, TN. S. A. David and J. M. Vitek. Pages 957-961. June,1992.

[323] H. E. Fujimura, E. Ide and H. Inoue. Estimation of contact tip-work piece distance in gas metal arc welding. Welding International. Volume 2(6), Pages 522-528, 1988.

[324] M. Kawahara. Tracking control system for complex shape of welding groove using image sensor. Proceedings of IFAC/IFIP Symp.-Real Time Digital Control Applications. Guadalajara, Mexico. Pages 257-263, Jan., 1983.

REFERENCES LIST FOR CHAPTER 3

[325] C. G. Morgan, J. S. E. Bromley, P. G. Davey and A. R. Vidler. Visual guidance techniques for robot arc welding. Proceedings of SPIE Third International Conference on Robot Vision and Sensory Controls RoViSeC3. Cambridge, MA. Volume 449, Pages 390-399, November, 1983.

[326] M. Dufour and G. Begin. Adaptive robotic welding using a rapid image pre-processor. Proceedings of Third International Conference on Robot Vision and Sensory Controls RoViseC3. Cambridge, MA. Volume 449, Pages 338-345. D. P. Cassasent and E. L. Hall. November, 1983.

[327] M. Dufour and P. Cielo. Optical inspection for adaptive welding. Application Optic. Volume 23(4), Pages 271-275, April, 1984.

[328] D. A. Dornfeld and M. Tomizuka. Development of a comprehensive control strategy for gas metal arc welding. Proceedings of 11th Conference Prod. Research and Technology. Pittsburgh,PA. Soc. of Manuf. Eng. Pages 271-275, May, 1984.

[329] J. E. Agapakis, K. Masubuchi and N. Wittels. General visual sensing techniques for automated welding fabrication. Proceedings of the 4th International Conference on Robot Vision and Sensory Controls RoViSeC4. London, UK. Pages 103-114, October, 1984.

[330] W. F. Clocksin, J. S. E. Bromley, P. G. Davey, A. R. Vidler and C. G. Morgan. An implementation of model-based visual feed-back for robot arc welding of thin sheet steel. International Journal of Robotics Research. Volume 4(1), Pages 13-26, 1985.

[331] N. R. Corby Jr. Machine vision algorithms for vision guided robotic welding. Proceedings of 4th International Conference on Robot Vision and Sensory Controls, RoViseC5. London, UK. Pages 137-147, October, 1985.

[332] Z. Smati, D. Yapp and C. J. Smith. Laser guidance system for robotics. Proc. of 4th Int. Conf. on Robot Vision and Sensory Controls RoViseC4. London, UK. Pages 91-101, October, 1984.

[333] N. Nayak, D. Thompson, A. Ray and A. Vavrek. Conceptual development of an adaptive real-time seam tracker for welding

automation. Proc. of 1987 IEEE Intl. Conf. on Robotics and Automation. Raleigh, NC. Pages 1019-1024, March, 1987.

[334] M. A. Wahab and M. J. Painter. Measurement and prediction of weld pool shape during gas metal arc welding using a noncontact laser profiling system. Proceedings of SPIE - The International Society for Optical Engineering. Singapore, December, Pages 34-39, 1996.

[335] E. L. Dereniak and G. Crowe. Optical Radiation Detectors. John Wiley & Sons. New York, NY, 1984.

[336] E. I. Romanenkov. Control of arc length on the basis of its spectral radiation. Welding Production. Volume23, Pages 53-54, 1976.

[337] C. A. Johnson and A. M. Sciaky. System for controlling length of welding arc. US Patent 3236997, 1966.

[338] R. B. Madigan, T. P. Quinn and T. A. Siewert. Sensing droplet detachment and electrode extension for gas metal arc welding. Proceedings of the 3rd International Conference on Trends in Welding Research. S. A. David and J. M. Vitek. Gatlinburg, TN. June, Pages 999-1008, 1992.

[339] Gary Bonser and A. Graham Parker. Robotic gas metal arc welding of small diameter saddle type joints using multistripe structured light. Optical Engineering. Volume 38(11), Pages 1943-1949, November, 1999.

[340] P. J. Li and Y. M. Zhang. Analysis of an arc light mechanism and its application in sensing of the GTAW process. Welding Journal (Miami, Fla). Volume 79(9), Pages 252s-260s, September, 2000.

[341] S. Liu, T. A. Siewert, G. A. Adam and H.G Lan. Arc welding process control from current and voltage signals. Proceedings of The Welding Institute Conference on Computer Technology in Welding. Pages 26-35, 1990.

[342] T. D. Manley and T. E. Doyle. Arc hydrogen monitoring for synergic GMAW Proceedings of the ASM 3rd International Conference on Trends in Welding Research. Gatlinburg, TN. S. A. David and J. M. Vitek. Pages 1027-1030, June, 1992.

REFERENCES LIST FOR CHAPTER 3

[343] T. A. Siewert, R. B. Madigan, T. P. Quinn and M. A. Mornis. Through-the-arc sensing for real-time measurement of gas metal arc weld quality. In Proceedings of International Conference on computerization of Welding Information IV, Orlando, FL. National Inst. of Standards and Technology MSEL. Gaithersburg, MD, Pages 198–206, 1992.

[344] T. A. Siewert, R. B. Madigan, T. P. Quinn and M. A. Mornis. Through-the-arc sensing for monitoring arc welding. Proceedings of the ASM 3rd International Conference on Trends in Welding Research. Gatlinburg, TN. S. A. David and J. M. Vitek. Pages 1037-1040, June, 1992.

[345] E. Murakami, K. Kugai and H. Yamamoto. H. Nomura. Dynamic analysis of arc length and its application to arc sensing. Sensors and Control Systems in Arc Welding. Chapman & Hall. London, UK. English Translation of the Original 1991 Japanese Edition. Chapter 25, Pages 216-227, 1994.

[346] L. A. Lott, J. A. Johnson and H. B. Smartt. Real-time ultrasonic sensing of arc welding processes. Proc. of 1983 Symp. on Nondestructive Evaluation Applications and Materials. Metals Park, OH. Pages 13=22, 1983.

[347] N. M. Carlson and J. A. Johnson. Ultrasonic inspection of partially completed weld using pattern-recognition techniques. Review of Progress in Quantitative Nondestructive Evaluation. D. O. Thompson and D. E. Chimenti. Plenum Publishing Co. New York, NY. Pages 773-780, 1986.

[348] J. A. Johnson, N. M. Carlson, R. T. Allemeier and D. G. Bannister. Ultrasonic and video computerized data acquisition for automated welding. Proceedings of Computer technology in welding. Cambridge, UK. June, 1988.

[349] J. A. Johnson, N. M. Carlson and L. A. Lott. Ultrasonic wave propagation in temperature gradients. Journal of Nondestructive Evaluation. Volume 6(3), Pages 147-157, 1988.

[350] P. Banerjee, J. Liu and B. A. Chin. Infrared thermography for non-destructive monitoring of weld penetration variations. Proceedings of ASME Japan/USA Symposium Flex. Auto. San Francisco, CA. July, Pages 291-295, 1992.

[351] V. J. Pavone. Univision 11, a vision system for arc welding robots. Proceedings of AWS Conference on Automation and Robotics For Welding. Pages 91-103, February 22-23, 1983.

[352] G. M. Graham and C. I. Ume. Laser array generated ultrasound for weld quality control. Proceedings of the ASM 4th International Conference on Trends in Welding Research. Gatlinburg, TN. H. B. Smartt, J. A. Johnson and S. A. David. June 5-8, Pages 677-681, 1995.

[353] P. Boughton, G. Rider and C. J. Smith. Feedback control of weld penetration. Proceedings of 4th International Conference on Advances in Welding Processes. Harrogate, England. Pages 203-215, May, 1978.

[354] C. N. Peters. The use of backface penetration control methods in synergetic pulsed MIG welding. Cranfield Institute of Technology. Cranfield, UK, 1986.

[355] D. Stone, J. S. Smith and J. Lucas. Sensor for automated weld bead penetration control. Measurement Science and Technology. Volume 1, Pages 1143–1148, 1990.

[356] J. B. Song and D. E. Hardt. A thermally based weld pool depth estimator for real-time control. Proceedings of 3rd International Conference on Trends in welding research. Gatlinburg, TN. Pages 975–980, June, 1992.

[357] S. I. Rokhlin. In-process radiographic evaluation of arc welding. Materials Evaluation. Volume 47, Pages 219-224, 1989.

[358] S. I. Rokhlin, K. Cho and A. C. Guu. Closed-loop process control of weld pool penetration using real-time radiography. Materials Evaluation. Volume 47, March, Pages 363-369, 1989.

[359] R. Fenn. Ultrasonic monitoring and control during arc welding. Welding Journal. Pages 18-22, September, 1985.

[360] J. A. Johnson and N. M. Carlson. Noncontact ultrasonic sensing of weld pools for automated welding. Proceedings of 3rd International Symp. on Nondestructive Characterization of Materials. Saarbrucken,FRG. Pages 854-861, 1988.

[361] N. M. Carlson , J. A. Johnson and D. C. Kunerth. Control of GMAW: Detection of discontinuities in the weld pool. Welding Journal. Volume 69(7), Pages 256s-263s, July, 1990.

[362] N. M. Carlson, J. A. Johnson, E. D. Larsen, A. Van Clark, S. R. Schaps and C. M. Fortunko. Ultrasonic sensing of GMAW Laser/EMAT defect detection system. International Conference on ASM 3rd Trends in Welding Research. Gatlinburg, TN. S. A. David and J. M. Vitek, June, Pages 859-863, 1992.

[363] J. M. Katz and D. E. Hardt. Ultrasonic measurement of weld penetration. Proceedings of Control of Manufacturing Process and Robot Systems. D. E. Hardt and W. J. Book. ASME, New York. Pages 79-95, November, 1983.

[364] D. E. Hardt and J. M. Katz. Ultrasonic measurements of weld pool penetration. Welding Journal. Volume 63, Pages 273s-281s, 1989.

[365] S. B. Zhang and Y. M. Zhang and R. Kovacevic. Noncontact ultrasonic sensing for seam tracking in arc welding processes. Journal of Manufacturing Science and Engineering, Transactions of the ASME. Volume 120(3), Pages 600-608, August, 1998.

[366] R. R. Stroud. Problems and observations whilst dynamically monitoring molten weld pools using ultra sound. British Journal of Non-Destructive Testing. Volume 31, January, Pages 29–32, 1989.

[367] S. Nagarajan, K. N. Groom and B. A.Chin. Infrared sensors for seam tracking in gas tungsten arc welding processes. Proceedings of the ASM 2nd International Conference on Trends in Welding

Research. Gatlinburg, TN. June. S. A. David and J. M. Vitek, Pages 951-955, 1989.

[368] H. E. Beardsley, Y. M. Zhang and R. Kovacevic. Infrared sensing of full penetration state in GTAW. Int. J. of Machine Tools and Manufacture. Vol 34, No 8, Pages 1079-1090, 1994.

[369] P. Banerjee, S. Govardhan, H. C. Wikle, J. Y. Liu and B. A. Chin. Infrared sensing for on-line weld geometry monitoring and control. ASME Transactions, Journal of Engineering for Industry. Volume 117, Pages 323-330, August, 1995.

[370] P. W. Ramsey, J. J. Chyle, J. N. Kuhr, P. S. Myers, M. Weiss and W. Groth. Infrared temperature sensing systems for automatic fusion welding. Welding Journal. Volume 42(8), Pages 337s-346s, August, 1963.

[371] J. L. Fihey, P. Cielo and G. Begin. On-line weld penetration measurement using an infrared sensor. Proceedings of International Conference Welding in Energy-Related Projects. Toronto, Canada. Pages 177-188, September, 1983.

[372] T. T. Lin, K. Groom, N. H. Madsen and B. A. Chin. Infrared sensing techniques for adaptive robotic welding. Proceedings of the Conference on Modeling and Control of Casting and welding Processes. Santa Barbara, CA, Pages 19-31, 1986.

[373] S. Nagarajan, W. H. Chen and B. A. Chin. Infrared sensing for adaptive arc welding. Welding Journal. Volume 68(11), Pages 462s-466s, 1989.

[374] K. N. Groom, S. Nagarajan and B. A. Chin. Automatic single V groove welding utilizing infrared images for error detection and correction. Welding Journal. Pages 441s-445s, December, 1990.

[375] S. M. Govardhan, H. C. Wikle, S. Nagarajan and B. A. Chin. Real-time welding process control using infrared sensing. Proceedings of the American Control Conference. Pages 1712-1716. Seattle, WA. June, 1995.

REFERENCES LIST FOR CHAPTER 3

[376] K. Masubuchi, D. E. Hardt, H. M. Paynter and W. Unkel. Improvement of reliability of welding by in-process sensing and control. Proceedings of Trends in Welding Research in The United State. New Orleans, LA. Pages 667-688, November16-18, 1981.

[377] D. M. Ainscough. Automatic Control of Weld Penetration. The University of Liverpool. Liverpool, UK, 1987.

[378] W. Lucas and R. S. Mallet. Automatic control of penetration in pulsed TIG welding. The Welding Institute. London, UK, 1975.

[379] D. J. Kotecki, D. L. Cheevr and D. G. Howden. Mechanism of ripple formation during weld solidification. Welding Journal. Volume 51, Pages 386s-391s, 1972.

[380] R. B. Madigan et al. Computer based control of full penetration GTA welding using pool oscillation sensing. Proceedings of a Conference on Computer Technology in Welding. The Welding Institute, London, UK, 1986.

[381] K. Andersen, G. E. Cook, R. J. Barnett and A. M. Strauss. Synchronous weld pool oscillation for monitoring and control. IEEE Transactions on Industry Applications. Volume 33(2), Pages 464-471, March/April, 1997.

[382] R. Best. Phase-Locked Loops: Theory, Design and Applications. Mc-Graw Hill. New York, NY, 1993.

[383] R. J. Barnett, G. E. Cook, J. D. Brooks and A. M. Strauss. A weld penetration control system using synchronized current pulses. Proceedings of the ASM 4th International Conference on Trends in Welding Research. Gatlinburg, TN. H. B. Smartt, J. A. Johnson and S. A. David. June 5-8, Pages 727-732, 1995.

[384] K. Andersen. Synchronous weld pool oscillation for monitoring and control. Vanderbilt University. Nashville, TN, 1993.

[385] R. J. Barnett. Sensor development for multi-parameter control of gas tungsten arc welding. Vanderbilt University. Nashville, TN, 1993.

[386] A. S. Tam and D. E. Hardt. Weld pool impedance for pool geometry measurement: stationary and nonstationary pools. ASME Transactions: Journal of Dynamic systems, Measurement and Control. Volume 111, December, Pages 545–553, 1989.

[387] R. T. Deam. Weld pool frequency: A new way to define a weld procedure. Proceedings of 2nd International Conference on Trends in Welding Research. Gatlinburg, TN. S. A. David and J. M. Vitek. Pages 967-971, May 18-22, 1989.

[388] G. Ouden, Y. H. Xiao and M. J. Hermans. The role of weld pool oscillation in arc welding. International Journal for the Joining of Materials. Volume 5(4), Pages 123-129, December, 1993.

[389] J. A. Johnson, N. M. Carlson and H. B. Smartt. Detection of metal-transfer mode in GMAW. Proceedings of ASM International conference on Recent Trends in Welding Science and Technology. EG&G Idaho, Inc. Idaho Falls. Pages 377-381, May, 1989.

[390] J. A. Johnson, N. M. Carlson, H. B. Smartt and D. E. Clark. Process control of GMAW: Sensing of metal transfer Mode. Welding Journal. Volume 70, Pages 91s-99s, April, 1991.

[391] Y. Arata, K. Inoue, K. Futamata and T. Toh. Investigation on welding arc sound-effect of welding method and welding condition of welding arc sound. Transactions of Japan Welding Research Institute (JWRI) . Volume 8(1), 1979.

[392] M. A. Matteson, R. A. Morris and D. Raines. An optimal artificial neural network for GMAW arc acoustic classification. Proceedings of ASM 3rd International Conference on Trends in Welding Research. Gatlinburg, TN. S. A. David and J. M. Vitek. Pages 1031-1035, June, 1992.

[393] P. R. Heald, R. B. Madigan, T. A. Siewert and S. Liu. Mapping the droplet transfer modes for an ERIOOS-1 GMAW electrode. Welding Journal. Volume 73, Pages 38s- 44s, February, 1993.

[394] P. Hauptmann. Sensors: Principles and Applications. Prentice Hall. Englewood Cliffs, NJ, 1991.

[395] Y. S. Kim and T. W. Eagar. Analysis of metal transfer in gas metal arc welding. Welding Journal. Volume 72(6), Pages 269s-278s. June, 1993.

[396] G. Adam and T. A. Siewert. Sensing of GMAW droplet transfer modes using an ER 100s-1 electrode. Welding Journal. Volume 69(3), Pages 103s-108s, March, 1990.

[397] G. Rider. On line measurement of weld pool surface size. Proceedings of Welding and Fabrication in the Nuclear Industry. London, UK, British Nuclear Energy Society. Pages 351-359, April, 1979.

[398] G. Rider. Control of weld pool size and position for automatic and robotic welding. Proceedings of the SPIE Third International Conference on Robot Vision and Sensory Controls RoViseC3. Cambridge, MA. Volume 449, Pages 381-389, November, 1983.

[399] L. M. Sweet, A. W. Case Jr, N. R. Corby and N. R. Kuchar. Closed-loop joint tracking, puddle centering and weld process control using an integrated weld torch vision system. Proceedings of the ASME Winter Annual Meeting on Control of Manufacturing Processes and Robotic Systems. Boston, MA. D. E. Hardt and W. J. Book. Pages 97–105, November, 1983.

[400] Y. Arata and K. Inoue. Automatic control of arc welding by monitoring the molten pool. Transactions of Japan Welding Research Institute (JWRI) . Volume 1(1), Pages 99s-113s, 1972.

[401] Y. Arata, K. Inoue, M. Morita and G. Kawasaki. Automatic control of arc welding (report V)-Application of digital picture processing technique to automatic control. Transaction of Japan Welding Research Institute (JWRI). Volume 5(1), Pages 77-85, 1976.

[402] J. P. Boillot, P. Cielo, G. Begin, C. Michel, M. Lessard, P. Fafard and D. Villemure. Adaptive welding by fiber optic thermographic sensing: An analysis of thermal and instrumental considerations. Welding Journal. Volume 64(7), Pages 209s-217s, July, 1985.

[403] W. M. McCampbell, G. E. Cook, L. E. Nordholt and G. J. Merrick. The development of a weld intelligence system. Welding Journal. Volume 45(3), Pages 139s-144s, March, 1966.

[404] R.S. Baheti. Vision processing and control of robotic arc welding system. Proceedings of the 24th IEEE Conference on Decision and Control. Pages 1022-1024, Ft. Lauderdale, FL. December, 1985.

[405] K. S. Boo and H. S. Cho. Determination of a temperature sensor location for monitoring weld pool size in GMAW. Welding Journal. Miami, FL. Volume 73(11), Pages 265s-271s, 1994.

[406] R. Kovacevic and Y. M. Zhang. Real-time image processing for monitoring of free weld pool surface. ASME Transactions, Journal of Manufacturing Science and Engineering for Industry. Volume 119(4), 1997.

[407] R. Kovacevic and Y. M. Zhang. On-line measurement of weld fusion zone state using weld pool image and neurofuzzy model. Proceedings of the 19 IEEE International Symposium on Intelligent Control. Dearborn, MI. Pages 307-312, September, 1996.

[408] Y. M. Zhang, R. Kovacevic and L. Wu. Sensitivity of front-face weld geometry in representing the full penetration. Proceedings of Institution of Mechanical Engineers, Part B: Journal of Engineering Manufacture. Volume 206(3), Pages 191-197, 1992.

[409] Y. M. Zhang, R. Kovacevic and L. Wu. Three-dimensional vision sensing weld penetration. Proceedings of the IASTED International Conference on Control and Robotics. Vancouver, Canada. Pages 301-304, August 4-6, 1992.

[410] Y. M. Zhang, B. L. Walcott and L. Wu. Dynamic modeling of full penetration process in GTAW. Proceedings of American Control Conference. Chicago. Volume 4, Pages 3345-3349, 1992.

[411] Y. M. Zhang, L. Wu, B. L. Walcott and D. H. Chen. Determining joint penetration in GTAW with vision sensing of weld-face geometry. Welding Journal. Volume 72, October, Pages 463s-469s, 1993.

REFERENCES LIST FOR CHAPTER 3

[412] Y. M. Zhang, H. E. Beardsley and R. Kovacevic. Real-time image process in 3D measurement of weld pool surface. ASME International Mechanical Engineering Congress. Volume 68, Pages 255-262, 1994.

[413] R. Kovacevic, Y. M. Zhang and S. Ruan. Three-dimensional measurement of weld pool surface. Proceedings of the International Conference on Model and Control of Joining Processes. Orlando FL. Pages 600-607, December, 1993.

[414] Y. M. Zhang, R. Kovacevic and L. Wu. On-line measure of full penetration weld geometry. Proceedings of the 12th World Congress of the IFAC. Sydney, Australia. Volume 8, Pages 97-100, July 18-23, 1993.

[415] R. Kovacevic and Y. M. Zhang. Machine vision recognition of weld pool in gas tungsten arc welding. Proceedings of Institution of Mechanical Engineers. Part B: Journal of Engineering Manufacture. Volume 209(B2), Pages 141-152, 1995.

[416] R. Kovacevic, Y. M. Zhang and S. Ruan. Sensing and control of weld pool geometry for automated GTA welding. ASME Transactions, Journal of Engineering for Industry. Volume 117(2), Pages 210-222, 1995.

[417] R. kovacevic and Y. M. Zhang. Weld pool sensing and control: 2D shape 3D surface. Proceedings of the International Symposium in Materials Science and Technology. Harbin, China. Pages 379-384, June 4-7, 1995.

[418] Y. M. Zhang, L. Li and R. Kovacevic. Dynamic correlation between weld shape and weld penetration. ASME International Mechanical Engineer Congress. ASME. San Francisco, CA. Volume 69, Pages 883-898, November 12-17, 1995.

[419] Y. M. Zhang, L. Li and R. Kovacevic. Monitoring of weld pool appearance for penetration control. Proceedings of the ASM 4th International Conference on Trends in Welding Research. Gatlinburg, TN. H. B. Smartt, J. A. Johnson and S. A. David. June 5-8, Pages 683-688, 1995.

[420] R. Kovacevic, Y. M. Zhang, E. Liguo and H. Beardsley. Dynamic analysis metal transfer process for GMAW control. ASM Journal of Engineering Materials and Technology. July, 1996.

[421] R. Kovacevic, Z. N. Cao and Y. M. Zhang. Role of welding parameters in determining the geometrical appearance of weld pool. ASME Transactions, Journal of Engineering Materials and Technology. Volume 118, Pages 589-596, October, 1996.

[422] R. Kovacevic, Y. M. Zhang, L. Li and H. Beardsley. Sensing and control weld geometrical appearance. Proceedings of the 26th Conference on Production Engineering. Budva, Yugoslavia. October, 1996.

[423] Y. M. Zhang and R. Kovacevic. Monitoring of three-dimensional arc weld surface. 26th Conference on Production Engineering. Budva, Yugoslavia. October 2-4, 1996.

[424] Y. M. Zhang, R. Kovacevic and L. Li. Characterization and real-time measurement of geometrical appearance of the weld pool. International Journal of Machine Tools and Manufacturing. Volume 36(7), Pages 799-816, 1996.

[425] Y. M. Zhang, Z. N. Cao and R. Kovacevic. Numerical analysis of fully penetrated weld pools in gas tungsten arc welding. Proceedings of Institution of Mechanical Engineers, Part C: Journal of Mechanical Engineering Science. Volume 210(2), Pages 187-195, 1996.

[426] R. Kovacevic and Yu M. Zhang. Neurofuzzy model-based weld fusion state estimation. IEEE Control Systems. Pages 30-42, April, 1997.

[427] Y. M. Zhang and R. Kovacevic. Real-time sensing of sag geometry during GTA welding. ASME Journal of Manufacturing Science and Engineering. Volume 119(2), Pages 1-10, May, 1997.

[428] Y. M. Zhang, L. Li and R. Kovacevic. Dynamic estimation of full penetration using geometry of adjacent weld pool. ASME journal of Manufacturing Science and Engineering. Volume 119(2), May, 1997.

[429] R. D. Richardson and R. W. Richardson. The measurement of two-dimensional arc weld pool geometry by image analysis. Proceedings of the ASME Winter Annual Meeting on Control of Manufacturing Processes and Robotic Systems. D. E. Hardt and W. J. Book. ASME. Boston, MA. Pages 137-148, November, 1983.

[430] C. W. Lee and S. J. Na. Study on the influence of reflected arc light on vision sensors for welding automation. Welding Journal. Miami, FL. Volume 75(12), Pages 379s-387s, December, 1996.

[431] G. Linden and G. Lindskog. A control system using optical sensing for metal-inert gas arc welding. Proceedings of TWI Conference. November, 1980.

[432] C. Conrardy. Control of GMAW with coaxial vision. Ohio State University, 1991.

[433] P. L. Taylor, A. D. Watkins, E. D. Larsen and H. B. Smartt. Integrated optical sensor for GMAW feedback control. ASM 3rd International conference on Trends in Welding Research. Gatlinburg, TN. S. A. David and J. M. Vitek. June, Pages 1049-1053, 1992.

[434] S. Subramaniam, D. R. White, D. J. Scholl, and W. H. Weber. In situ optical measurements of liquid drop tension in gas metal arc welding. Journal of Physics D: Applied Physics. Volume 31(16), Pages 1963-1967, August, 1998.

[435] C. W. Lee and S. J. Na. Study on the influence of reflected arc light on vision sensors for welding automation. Welding Journal (Miami, Florida), Vol 75, No 12, Pages 379s-387s. December, 1996.

[436] R. C. Anderson. Inspection of Metals: Visual Examination. American Society of Metals. Metals Park, OH, 1993.

[437] K. Andersen, R. J. Barnett, J. F. Springfield and G. E. Cook. Weldsmart: A Vision-Based System for Quality Control. Vanderbilt University. NASA Contract NAS8-37685, 1992.

[438] R. J. Barnett, G. E. Cook, A. M. Strauss, K. Anderson and J. F. Springfield. A vision-based weld quality evaluation system. Proceedings of the ASM 4th International Conference on Trends in Welding Research. Galtinburg, TN. H. B. Smartt, J. A. Johnson and S. A. David. June 5-8, Pages 689-694, 1995.

[439] G. E. Cook, A. M. Wells Jr, H. M. Floyd and R. L. McKeown. Analyzing arc welding signals with a micro-computer. IEEE IAS Annual Meeting. Pages 1282-1288. San Francisco, CA. October, 1982.

[440] J. C. Papritan and S. C. Helzer. Statistical process control for welding. Welding Journal. Pages 44-48, March, 1991.

[441] H. R. Castner and D. Barborak. Expert system for diagnosis of discontinuities in gas metal arc welds. Edison Welding Institute. Columbus, OH. January, MR9101, 1991.

[442] J.-S. R. Jang, C.-T Sun, and E. Mizutani. Neuro-Fuzzy and Soft Computing. Prentice Hall, Upper Saddle River, NJ, 1997.

[443] Y. M. Zhang, R. Kovacevic and L. Wu. Dynamic analysis and identification of gas tungsten arc welding process for full penetration control.

Transactions of ASME, Journal of Engineering for Industry. Volume 118, February, Pages 123-136, 1996.

[444] Y. M. Zhang, R. Kovacevic and L. Li. Adaptive control of full penetration gas tungsten arc welding. IEEE Transaction on Control Systems Technology. Volume 4(4), Pages 394-403, July, 1996.

[445] A. Rock, X. Xu and J. E. Jones. Investigation of an artificial neural system for a computerized welding vision system. Proc. of the ASM 2nd Int. Conf. on Trends in Welding Research. Gatlinburg, TN. editors S. A. David and J. M. Vitek. Pages 957-965, May, 1989.

[446] R. Kovacevic, Y. M. Zhang and H. E. Beardsley. Weld penetration control with infrared sensing and neural network technology.

Proceedings of the International Conference on Modeling and Control of Joining Processes. Orlando, FL. Pages 393-400, December, 1993.

[447] R. W. Richardson, w. A. Penix and K. L. Boyer. Interpretation of arc weld images by vision analysis. Japan/USA Symposium on Flexible Automation. Pages 309-312, 1992.

[448] H. Onda, Y. Nishinaga and K. Ono. Welding defect identification by artificial neural networks. Japan/USA Symposium on Flexible Automation. Pages 313-316, 1992.

[449] R. Du, M. A. Elebestawi and S. M. Wu. Automated monitoring of manufacturing processes, Part 1: monitoring methods. Transactions of ASME, Journal of Engineering for Industry. Volume 117, Pages 121-132, May, 1995.

[450] R. Du, M. A. Elebestawi and S. M. Wu. Automated monitoring of manufacturing processes, Part 2: applications. Transactions of ASME, Journal of Engineering for Industry. Volume 117, Pages 133-1141, May, 1995.

[451] M. Dufour, X. Maldague and P. Cielo. Environmental-noise analysis in active-vision systems for adaptive welding. Proceedings of the SPIE, Optical Techniques for Industrial Inspection. Quebec City, Quebec, Canada. Volume 665, Pages 321-332, June, 1986.

[452] T. Araya and S. Saikawa. Recent activities on sensing and adaptive control of arc welding. Proceedings of the 3rd International Conference on Trends in Welding Research. S. A. David and J. M. Vitek. Gatlinburg, TN. June, Pages 833-842, 1992.

[453] C. D. Yoo, K. I. Koh and H. K. Sunwoo. Investigation on arc light intensity in gas metal arc welding. Part 2: Application to weld seam tracking. Proceedings of the Institution of Mechanical Engineers.. London, England. Pages 355-363, 1997.

[454] T. P. Quinn, C. Smith, C. N. McCowan, E. Blackhowaik and R. B. Madigan. Arc Sensing for defects in constant-voltage Gas metal arc welding. Welding Journal. Pages 322s-328s. September, 1999.

[455] D. R. Blackmon and F. W.Kearney. A real-time quality approach to quality control in welding. Welding Journal. August, 1983.

[456] J. F. Justice. Sensors for robotic arc welding. Proceedings on AWS Conference on Automation and Robotics For Welding. Pages 203-210, Indianapolis, IN. February, 1983.

[457] G. C. Cook, K. Anderson and R. J. Barrett. Feedback and adaptive control in welding. Proceedings of the 2nd International Conference on Trends in Welding Research. Gatlinburg, TN. S. A. David and J. M. Vitek. Key Note Address. Pages 891-903, May, 1989.

[458] R. W. Gellie. Sensing for automated welding. Proceedings of the 31st Annual Conference on welding and Computers. Sydney, Australia. Pages 193-200, October, 1983.

[459] J. W. Kim and S. J. Na. A study on an arc sensor for gas metal arc welding horizontal fillets. Welding Journal. Pages 216s-221s, August, 1991.

[460] J. W. Kim and S. J. Na. Study on arc sensor algorithm for weld seam tracking in gas metal arc welding butt joints. Proceedings of the Institution of Mechanical Engineers, Part B:Journal of ISSN. Volume 205(B4), Pages 247-255, 1991.

[461] Y. Li, L. Wu, D. Cheng and J. E. Middle. Machine vision analysis of welding region and its application to seam tracking in GTAM and GMAW. Proc. of the 3rd Intl. Conf. on Trends in Welding Research. Gatlinburg, TN. S. A. David and J. M. Vitek. June, Pages 1021-1025, 1992.

[462] H. Nomura. Sensors and Control Systems in Arc Welding. Chapman & Hall. London, UK. English Translation of the Original 1991 Japanese Edition, 1994.

[463] H. Nomura. Analysis of questionnaire results on sensor applications to welding processes. Sensors and Control Systems in Arc Welding. Chapman & Hall. London, UK. English Translation of the Original 1991 Japanese Edition. Chapter 6, Pages 76-85, 1994.

[464] H. B. Smartt, J. A. Johnson and S. A. David. Proceedings of the ASM 4rd International Conference on Trends in Welding Research. ASM International. Gatlinburg, TN. June, 1995.

[465] H. Nomura. Future trends. Sensors and Control Systems in Arc Welding. Chapman & Hall. London, UK. English Translation of the Original 1991 Japanese Edition. Chapter 7, Pages 86-87, 1994.

[466] S. B. Chen, L.Wu, Q. L. Wang and Y. C. Liu. Self - learning fuzzy neural networks and computer vision for control of pulsed GTAW. Welding Journal. Pages 201s-209s, May, 1997.

[467] R. Reilly. Real-time weld quality monitor control GMAwelding. Welding Journal. Pages 36-41, March, 1991.

Chapter 4

Gas Metal Arc Welding: Automatic Control

In this chapter, we survey the automatic control techniques that have been used in the Gas Metal Arc Welding (GMAW) process. Basically, in welding operation there is *manual* control and *automatic* control. We begin with a brief disscusion of manual control. Then we discuss in detail many of the automatic control techniques for welding that have been reported in the literature.

4.1 Automatic Welding

Traditionally, welding processes have been manually operated and are based on trial and error and/or on operator's experience. When improved control is required under manual control, an automatic *welding procedure or control* is established [468].

Automatic welding, simply means that some aspects of the welding operation is performed without the intervention of human such as welder or welding operator. In most of automatic welding operations, a welder is required to make initial preparations and then monitors the overall operation. The advantages of an automatic welding system include the following:

1. Consistency in welding quality.

2. Increased welding production and consequent reduction of production costs.

3. Integration with other automatic operations of the industry.

4. Absence of human fatigue and/or error.

5. Absence of loss of human life in case of severe accidents.

However, automatic welding also has some disadvantages such as listed below:

1. Extensive planning of procedural steps.

2. Higher capital investments leading to uneconomic investment for small operations.

Robots play an important part in the automation of welding processes, particularly for arc welding processes such as GMAW. Robots are extensively used in industries such as automobile manufacturing.

According to Cook, et al. [469], the goals of feedback control in welding are:

1. Producing welds with desired mechanical and metallurgical properties.

2. Controlling the microstructure during solidification and cooling.

3. Sensing and controlling discontinuous formations to acceptable levels.

There are two basic techniques for controlling any process.

1. The first one is *open-loop* control, where the process is driven by an *input* to correspond to the desired output, with the hope that the *output* will be the desired one.

2. The second technique is an extension of the open loop control approach, where we *measure or sense* the output variables and then *feed* them *back* to compare with the desired (reference) variables, detect the error between them, and then adjust (or control) the inputs to the process until the actual output variables exactly match the desired output variables. This approach is called *automatic control*, *closed-loop control*, or *feedback control*.

4.1. AUTOMATIC WELDING

A simplified diagram showing both open-loop and closed-loop control systems is shown in Figure 4.1. Under closed-loop control, we also have

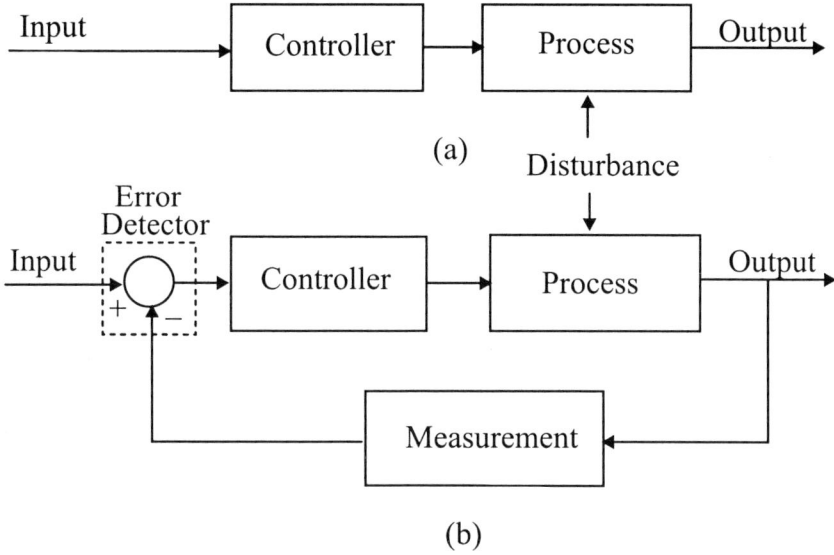

Figure 4.1: (a) Open-loop and (b) Closed-loop control systems.

feed forward control as shown in Figure 4.2 [470].

There are several forms of closed-loop control strategies, such as optimal control [471], adaptive control [472], robust control [473], and learning or intelligent control [474]. These will be briefly described at the proper places in this chapter.

We notice from Figure 4.1 that closed-loop control requires that the feedback signals are generated by employing sensors or transducers to convert the physical parameters into electrical signals. For simulation purposes, one also has to obtain a mathematical *model* in terms of the *input* and *output* variables of the process.

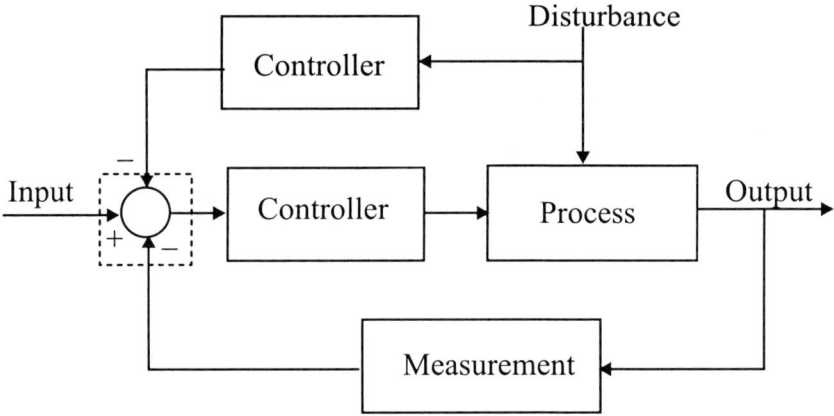

Figure 4.2: Feedforward control system.

4.2 Control of Process Variables

4.2.1 Arc Length Control

Control of arc length in the welding process is important to ensure consistent heat input, constant melting rate, and stable performance of the process. In particular, arc length determines the transfer mode, arc stability, and the deposition rate [475].

In GMAW, one simple way of controlling the arc length is to control the arc voltage. With a constant-current power source and variable wire-feed speed (WFS), the arc voltage (i.e., the process voltage) is used to drive the wire feed motor which in turn changes the arc length. On the other hand, with a constant-voltage power source and constant WFS, the changes in current are such as to provide a constant arc voltage (i.e., arc length) [475, 476, 477].

Using an arc length sensor, a closed-loop controller was developed for spray transfer GMAW process in [478]. A simple PID controller was designed where the coefficients of PID controller were determined from the process characteristics determined experimentally (see Figure 4.3, where CTWD refers to contact tip-to-workpiece distance). Based on a linearized model of the arc dynamics, a continuous-time arc current controller was designed and discretized for computer implementation [479].

4.2. CONTROL OF PROCESS VARIABLES

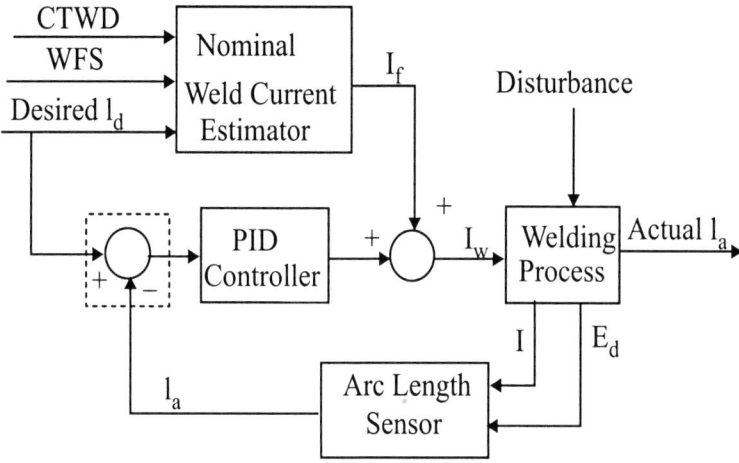

Figure 4.3: PID control system for arc length regulation.

4.2.2 Control of Mass and Heat Transfer

Here, we discuss control of mass and heat transfer, including control by pulsed current. The control of mass (metal) and/or heat transfer during the GMAW process is a very important consideration in employing automatic control schemes. For example, an increase of wire feed speed causes the increase of weld current which in turn increases the heat produced and hence the melting rate. Thus, it is very difficult to decouple the control of heat transfer from that of the mass transfer. However, the problem can be tackled by using the pulsed (current) control [480, 481].

In another investigation, Smartt and Einerson [482] considered a steady-state model for heat (H) and metal (G) transfer from the electrode to the workpiece in GMAW process. Using the relations between G and H, a PI-based control system was developed for maintaining the desired G and H by regulating the current.

By using pulsed current, desired characteristics of spray and dip transfer are obtained. A significant contribution to the advancement of pulsed GMAW technology is believed to be available through a system developed by the Welding Institute of Canada [483, 484]. A system where the pulsed current is generated by using arc voltage instead of wire-feed speed is shown in Figure 4.4. Essentially, the arc voltage

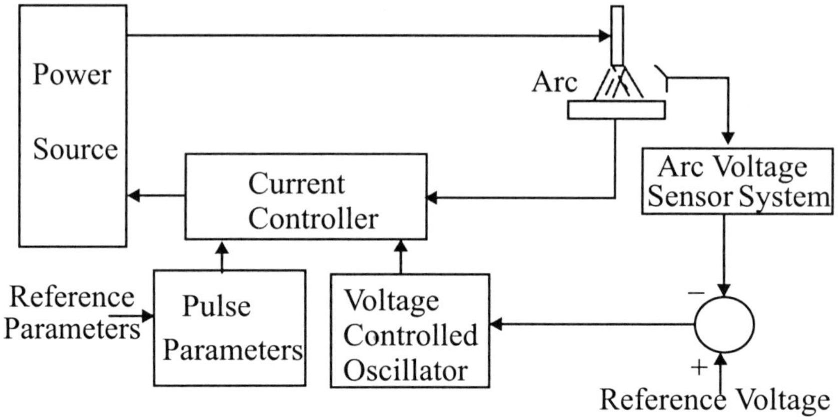

Figure 4.4: Frequency Modulated Pulse Current Feedback System

feedback signals control the pulse frequency to maintain a stable arc condition. The frequency modulated (FM) method of pulsing the current seems to be better than other methods because the parameters of the pulse (high current, low current, pulse duration) remain constant and hence result in regular and consistent metal transfer.

A synchro-pulse GMA method of pulsed power welding has been reported in [485]. The arc length is held constant by using arc voltage as the reference for a feedback control system.

4.2.3 Control of Weld Temperature and/or Cooling Rate

A temperature control system was designed for a consumable electrode GMAW process using an inexpensive optical system for measuring weld temperatures [486]. Dorfeld et al. [487] addressed the problem of controlling the temperature on the back side of weld plate using infrared thermography. In another work [488], a PID controller was designed to control the cooling rate.

An intelligent control system using both neural networks and fuzzy logic was developed by Einerson, et al. [489], for cooling rate and fill control. See also [490] for other works on controlling the temperature.

4.2. CONTROL OF PROCESS VARIABLES

4.2.4 Control of Weld Pool and its Geometry

A distributed source conduction model was presented in [491] for prediction and control of weld geometry with a real-time calibration of the model as shown in Figure 4.5, was discussed. Here D_d, W_d are the

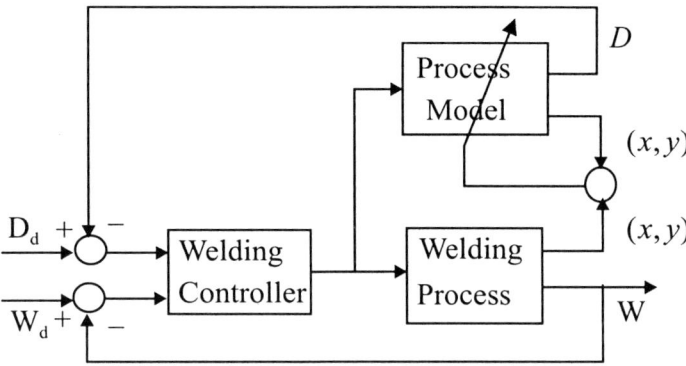

Figure 4.5: Schematic for a weld pool geometry (width and depth) control system.

desired and actual depth and width, respectively, of the weld pool and $\hat{\theta}(x,y), \theta(x,y)$ are the estimated and actual temperature distributions, respectively.

A feedback control system for weld penetration control based on the weld oscillation mode was developed for a fully automatic GTAW unit [492]. In this system, programming and implementation were carried out using the LabVIEW software of National Instruments [493]. Such a system is called *in-process penetration control* (IPPC), and the system was implemented for orbital tube welding.

An experimental facility was built for controlling the puddle geometry using a pseudo-gradient adaptive algorithm for self-tuning a PI-based puddle width controller for a consumable electrode GMAW process [494, 495, 496, 497]. A microprocessor-based control system was developed to join sheet metal parts for a GTAW process [498, 499], where the desired puddle area and puddle width are achieved by feeding back the measured area and width using a vision system. For the GMAW process, a multi-input (wire-feed rate and travel speed) and multi-output (weld bead geometry: width, depth and reinforcement) model was used in a scheduled-gain multivariable controller [500].

Using front-end (or side) infrared sensors, an on-line scheme for monitoring and controlling the weld geometry was presented for both GTAW and GMAW processes by Banerjee et al. [501]. In particular, the applicability of infrared thermography in sensing variations in bead width and depth of penetration due to variations in plate thickness, shielding gas composition, and minor element content was demonstrated by experimental verification.

4.2.5 Other Works on Control of Weld Pool Geometry

One of the earliest studies on methods for full penetration sensing and the use of steady-state puddle depression measurements in a closed-loop control experiment was reported in [503]. It suggests the use of smart or intelligent welding machines consisting of sensors, actuators, artificial intelligence, and automatic control.

A study of a welding control system, which used a line scan camera focused on the molten weld puddle to provide puddle width, along with an analog computer, was used to adjust electrode holder speed to maintain constant puddle width, is presented in [502]. A vision based system for control of GTAW pool width and, hence, weld penetration, was used in [504], with a real-time proportional control algorithm. A model-based visual feedback control system for detecting various welding parameters was designed in [505]. A backface penetration control system for DC pulsed TIG welding, utilizing a coherent optical bundle to transmit the image of backface bead to a video camera, was developed in [506].

4.2.6 Control of Droplet Transfer Frequency

It is desirable to develop a method of measuring droplet transfer mode and its frequency to monitor and control the GMAW welding process and achieve a desirable weld quality [507].

With droplet transfer frequency determined by arc light sensing, a closed-loop PID controller was developed for the GMAW process under spray transfer mode [478](see Figure 4.6). In this figure, $K_P, K_I,$ and K_D are proportional, integral and derivative constants (tuned empirically) of the PID controller, f_D and f_o are the *desired* and actual

4.2. CONTROL OF PROCESS VARIABLES

droplet transfer frequencies, and E_D is the output of the arc light detector. In particular, the droplet frequency was obtained by finding

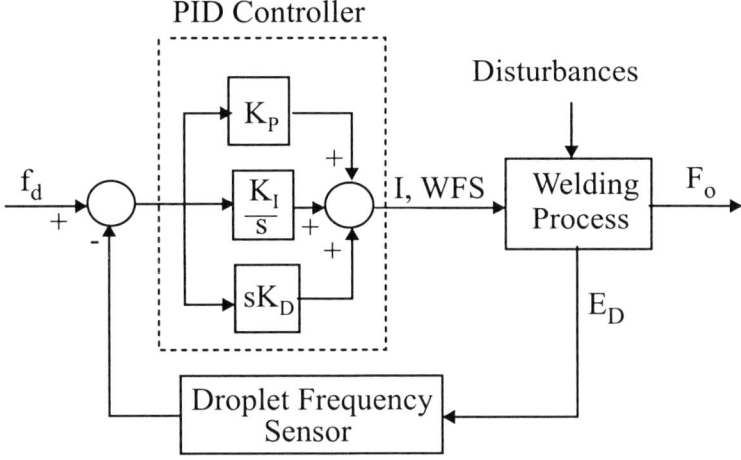

Figure 4.6: PID control system for droplet frequency regulation.

the frequency at which the maximum amplitude occurs in the power spectrum density (PSD) of the sensor (detector) frequency response.

At the Colorado School of Mines, a complete equipment instrumentation system was set up for experimental investigations on controlling the GMAW process using arc length and droplet detachment frequency. A typical welding equipment and instrumentation assembly with a constant current power source for droplet detachment frequency measurement is shown in Figure 4.7 [478]. In estimating droplet detachment frequency using an optical sensor, Madigan [478] measured voltage arc current, process voltage, wire feed speed, and detector voltage. These signals, sampled at 3000 Hz by the data acquisition computer, are passed through a low-pass filter with a cut-off frequency of 1000 Hz, which is sufficiently below the Nyquist frequency of 500 Hz to avoid aliasing [508]. Further, these signals are evaluated using power spectral density in order to express the time-domain data in terms of the frequency-domain data for analysis [509].

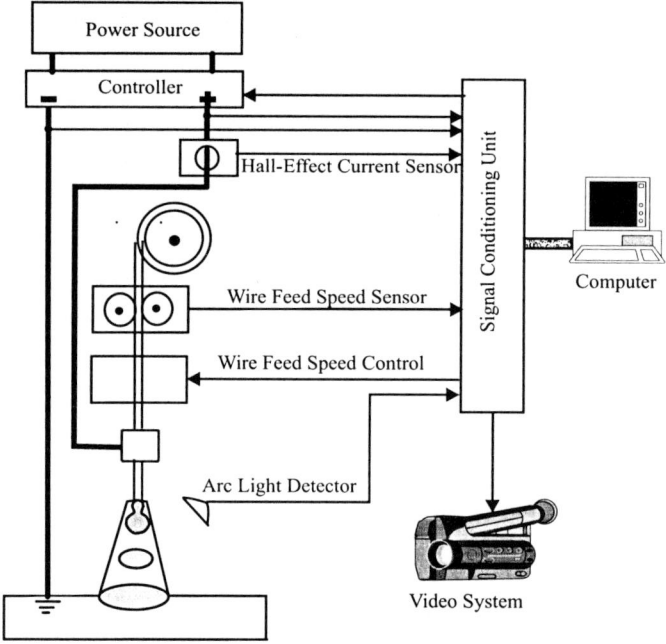

Figure 4.7: Welding equipment and instrumentation for experimental facility.

4.2.7 Control of Weld Penetration

Control of weld penetration is one of the most challenging problems in arc welding. Here the objective may be to maintain a constant weld penetration along the joint in spite of irregularities in the joint geometry. For full penetration, there are a number of control schemes suggested using a variety of measuring techniques, such as the back-face method [510, 511] or by measuring back-bead width [512, 513].

By locating a thermocouple near the weld pool to measure the temperature, a feedback signal proportional to weld penetration was generated in a control system for maintaining uniform penetration with torch speed as the control variable [514].

Madsen and Chin [515] used an IR sensor to predict the depth of penetration by scanning the weld pool in a direction transverse to torch travel. A correlation between the depth of penetration and the solid-

4.2. CONTROL OF PROCESS VARIABLES

ification time of the weld pool was found in [516]. Using a functional relationship between weld penetration, weld pool width, current, and torch speed, Rider measured the weld pool width to control the penetration by controlling the welding current and torch speed [517, 518]. Further, measuring the weld pool width and comparing it to the desired width, a feedback control system was reported in [519] for joints having fixed weld geometry. Also, using front-face ultrasonic measurements of the weld pool, a real-time weld penetration control system was suggested [520, 521].

Andersen, et al. [522], proposed a penetration control system based on the weld pool natural frequency of oscillation using a phase-locked loop (PLL), where the actual weld pool frequency is synchronized with the PLL, which locks and tracks the natural frequency of weld pool, being a function of the pool mass and hence, indirectly, of the pool geometry [523].

Some other works on this topic are listed. A method of calculating the dependence of the radiation of the controlled section of the weld pool on its geometrical parameters, without backing strip and full penetration, was discussed in [524]. A weld penetration control system was designed in [520], in which ultrasonic signals were employed to measure weld penetration by placing the transducers along the side of the molten weld pool. In [525], a feedback control system was designed to assure constant heat input to the workpiece, which resulted in more consistent weld penetration. Further works can be found in [511, 526].

4.2.8 Control of Joint Profile (Fill Rate) and Trajectory

A control system for controlling the joint profile is given in [527]. The trajectory controller is meant for providing the torch orientation and its path, where the orientation is defined by the longitudinal angles and the path is the change of positional coordinates with respect to time. In the design of a typical trajectory controller, the control law is obtained as [528]

$$u_t = f_1(G) + f_2(W_p) \quad (4.1)$$

where u_t is the torch lateral position, G is the joint profile geometry, W_p is the weld pool center position and f_1 and f_2 are functional relations, provided by the welding data base in accordance with the type of joint.

A simple seam tracking controller ($f_2 = 0$) is given by Khosla et al. [529].

Einerson, et al. [489] developed a strategy for GMAW for controlling the reinforcement (mass deposited) and the weld bead centerline cooling rate. The strategy involves the measurement of the weld joint transverse cross-sectional area *ahead* of the welding torch and the weld bead centerline cooling rate *behind* the weld pool, using a video camera. Further, the control scheme employed an intelligent component in terms of a combination of a neural network for controlling electrode speed and torch speed and a fuzzy logic algorithm for controlling the reinforcement G and the heat input H (see Figure 4.8, where R and S refer to torch speed and wire feed speed respectively).

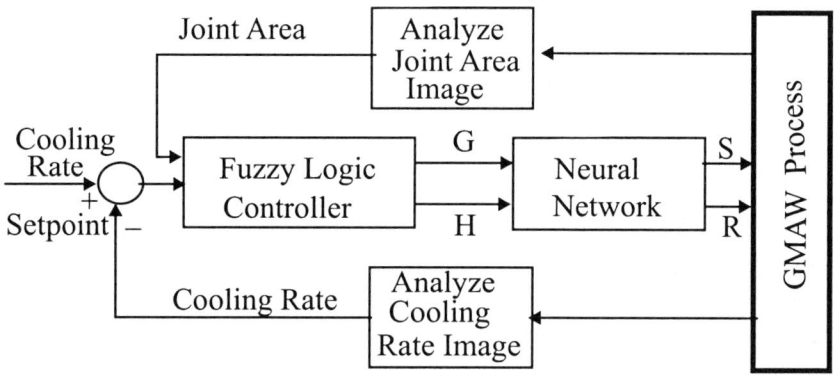

Figure 4.8: Intelligent control of the GMAW process.

A noncontact TV tracking system was built in [530]. It was suitable for flat horizontal welding of box section fabrications, incorporating video processing features specially suitable for real-time adaptive control of position and welding parameters.

A coaxial welding viewing system to accomplish automatic weld joint tracking for the GTAW process was developed and tested in [531, 504]. Richardson has performed the same work for the GMAW process [532]. A joint-tracking system was developed in [533, 534] using a new formulation for contact-tip-to workpiece distance. The system operates not only on welding current, but also on welding voltage and wire feed speed. Tomizuka in [535] gives a general exposition of the design

4.2. CONTROL OF PROCESS VARIABLES

of digital tracking controllers for manufacturing applications such as machining and welding.

4.2.9 Control of Other Variables or Conditions

In this subsection, we survey items not covered in the other sections. In a keynote address, Araya and Saikawa [536], gave an excellent account of activities (particularly in Japan) in sensing and adaptive control of arc welding. Sometimes, an automatic control system may also be required to detect any shortage conditions such as that of electrode and shielding gases [537]

In many industrial applications using robots, such as filling wide joints, arc welding requires the weaving of torch in a particular pattern (square, sinusoidal, sawtooth) depending upon the geometry of the weld [538, 539]. Early studies on control of joint configuration, control of weld line position, and control of molten pool condition can be found in [540],[541], and [542], respectively.

An active metal transfer control by monitoring excited droplet oscillations was presented by [543]. Controlled metal transfer implies controllable heat and mass inputs and improved weld quality. A combined primary and secondary power supply has been developed in [544] for gas-shielded metal-arc welding with a pulsed arc. The process control system incorporated additional functions of pulse control.

Tao in [545] conducted an experimental study to assess the impact of the weld power source control on feedback values of welding current and voltage relevant to seam tracking. In [546], it was noted that a low-cost, non-intrusive sensing technique, known as through-the-arc sensing, involves collecting and analyzing welding current and voltage signals.

An efficient method of identifying power supply pulsing parameters was developed in [547] for pulsed GMAW, based on statistical experimental design. Cullison in [548] notes that spatter is a result of unstable conditions in the arc. One way to reduce this problem is to stabilize the arc.

In [549], the results of an investigation dealing with the short gas metal arc welding with the emphasis on process stability are presented. Welding runs were made under different conditions and during each run the different process parameters were continuously monitored.

4.3 Classical Control: PI, PID and Others

Here, we review the GMAW process being controlled by classical techniques, particularly proportional-integral-derivative (PID) control.

One of the earliest applications of automatic control in a welding process is found in [514], which described the development of a servo (automatic) control system using feedback signals from temperature (thermocouple) measurements to maintain a constant weld penetration by adjusting the weld travel speed. For historical significance, the corresponding feedback control system is shown in Figure 4.9.

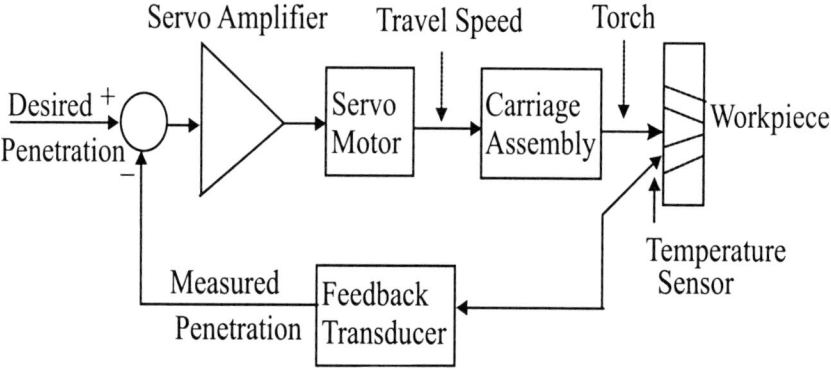

Figure 4.9: An early feedback control system for a welding process.

Although the title of the paper indicates "intelligence" as part of the system, there is no reference to any modern terminology, such as artificial neural networks, fuzzy logic, genetic algorithms, AI, or expert systems. Of course, these fields were not formally in existence at that time. Perhaps during those days, the action "automatic" was enough to qualify as an intelligent action!

A simple combined (voltage and current) control was advised for DC arc welding in [550, 551]. Here, two controllers were developed, one is based on a *quadratic* power-current relation and the other controller is based on an approximate *linear* relation.

A proportional-integral (PI) controller was used by Smartt and Einerson [482] to achieve a desired heat H and metal G transfer from the electrode to the workpiece in a GMAW process with spray transfer mode. The difference between a welding current based on model and

4.4. MULTIVARIABLE CONTROL

that of the actual measured current was used as a feedback to obtain correct wire feed speed and torch speed as shown in Figure 4.10 where R and S are the travel speed and wire feed speed, respectively, and I and I_m are the model and measured currents, respectively. Also, see

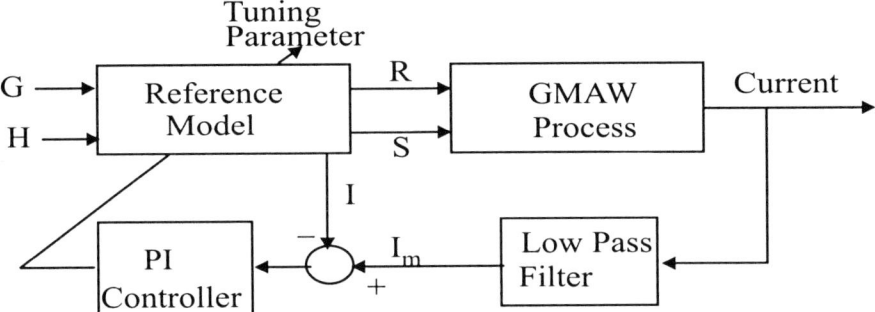

Figure 4.10: PI control of the GMAW process.

other relevant works by this group [552, 553, 554, 555, 556].

A 3-dimensional positional control system is discussed in [557] for both GTAW and GMAW that permits automatic correction for deviations in the actual weld path trajectory from preprogrammed anticipated trajectories. An international project involving different Scandinavian (Sweden, Denmark, Norway and Finland) Institutes to develop a control system using an optical sensor (photodiode array camera) for measurements of the groove geometry in front of the welding head was presented in [558].

Farson et al. [488] gives the design and simulation of a PID controller to control the cooling rate of a GMAW system. A simple feedback control system for a GMAW process was designed in [559], where the process was modeled as a first-order system with input as wire feed rate and output as arc length (arc voltage).

4.4 Multivariable Control

Hardt in [560] addressed the multivariable feedback control system to control the five output variables: weld geometry variables (width, depth and height) and thermal properties (CR and HZ), as shown in Figure 4.11 Here, HAZ is heat-affected zone, CR is the maximum cooling

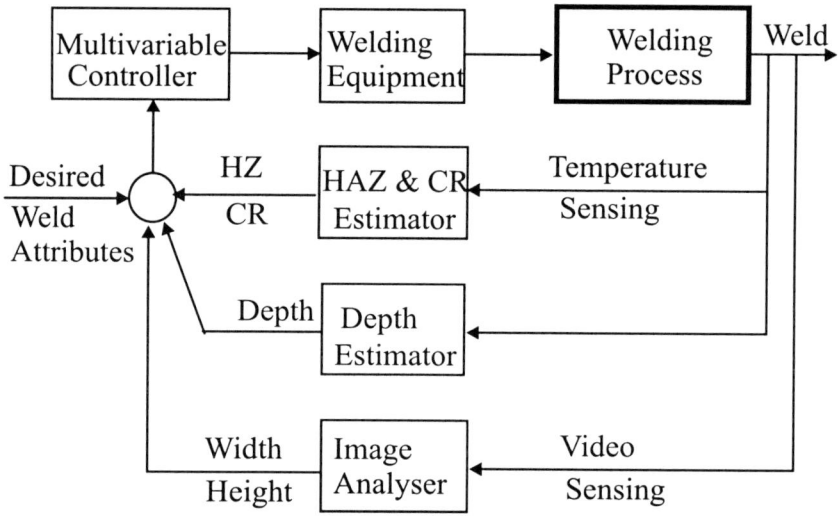

Figure 4.11: A multivariable feedback control system for the GMAW process.

rate. Also, see [561] for further discussion on a multivariable control system approach to GMAW. The work in [562] presents a similar treatment except the outputs to be controlled are bead width and depth. In another investigation on multivariable control of welding processes, Hardt and his associates [563, 564, 565] showed that high-frequency (3 to 10 Hz) weaving changes the temperature distribution in the weldment and significantly reduces the coupling between the desired outputs, the weld pool width, and the heat-affected zone width.

For other related work, see [566] for investigations on both single-input, single out-put (SISO) and multi-input, multi-output (MIMO) adaptive control schemes for a GMAW process using a discrete-time transfer function model that takes the inherent time delays in the process into account. Also, see related works [567, 568, 569, 562, 570].

In another multivariable control framework for GMAW process, Huissoon, et al. [571, 572], present a traditional guidance and control technique using a linearized model (based on analysis of small deviations from the nominal) of the original (nominal) nonlinear system and

4.5. OPTIMIZATION AND OPTIMAL CONTROL

then controlling the system around the nominal states or conditions [573]. Here, the original GMAW process is a nonlinear system, and the

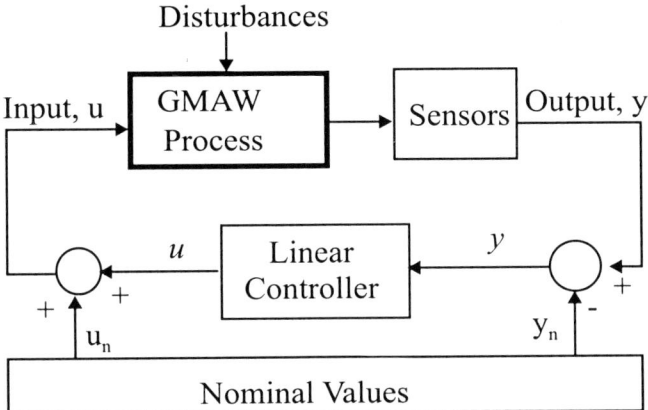

Figure 4.12: A multivariable linearized feedback controller for the GMAW process.

linearized system is obtained for the purpose of controller design.

Also, see [574] for a presentation of a multivariable linear controller designed to regulate the width and throat thickness of fillet welds during a GMAW process by simultaneously manipulating torch travel speed, power supply voltage, and wire-feed rate to achieve desired weld geometry. In this work the controller was designed using an empirically-derived linearized model of the welding process operating at a pre-selected operating point and using optimal control theory to ensure reference tracking, disturbance rejection, and robustness.

4.5 Optimization and Optimal Control

In a typical optimal control problem, we have a process or plant described by a differential (for a continuous system) or difference (for a discrete, digital, or computer controlled system) state equation and a performance criterion such as minimizing an error and/or the control effort [471]. In [494], both single-input, single-output (SISO) and multi-input, multi-output (MIMO) models of puddle geometry for a GMAW process are derived using off-line identification techniques such as recursive least-square identification and recursive maximum likelihood estimation [575].

In [576], a control strategy for two-axis welding torch positioning and velocity control for a typical GMAW process was developed using LQ control methods. The LQ design was compared with conventional PID controllers and it was found, via simulation, that high quality seam tracking could be accomplished with the LQ control strategy. An optimization analysis was carried out to find the optimal welding variables (groove area, heat input rate, and heat input per cm of weld length) for the minium residual stresses due to welding [577].

A simple mathematical method for the estimation of the optimum heat inputs in arc welding was developed in [578]. The optimization problem is to make the temperature field coincide with the required field during welding and is expressed as a quadratic function of heat inputs.

An experimental approach to selection of pulsing parameters in pulsed GMAW was proposed in [547]. An efficient method of identifying power supply pulsing parameters for pulsed GMAW based on statistical experimental design is presented.

4.6 Adaptive Control

First of all, let us note that in the control community the term *adaptive control* is used in the literature [472] to mean that the controller is designed so as to *adapt* for parameter variations and disturbances in the process. In this context, an adaptive controller is a controller with adjustable parameters and a mechanism for adjusting the parameters on its own (self-regulation or self-adjustment). But in the welding community, the term adaptive control is used somewhat loosely to mean that the process can adapt to the changing welding conditions, which is nothing but feedback control [469]. Here we take the former meaning.

A simple block diagram of the principle of adaptive control, in particular that of model reference adaptive control (MRAC), is shown in Figure 4.13 [579]. The idea is that the adaptive controller drives the physical process to follow the reference model over a range of parameters of the plant, hence it is called the *model reference adaptive control.*

4.6. ADAPTIVE CONTROL

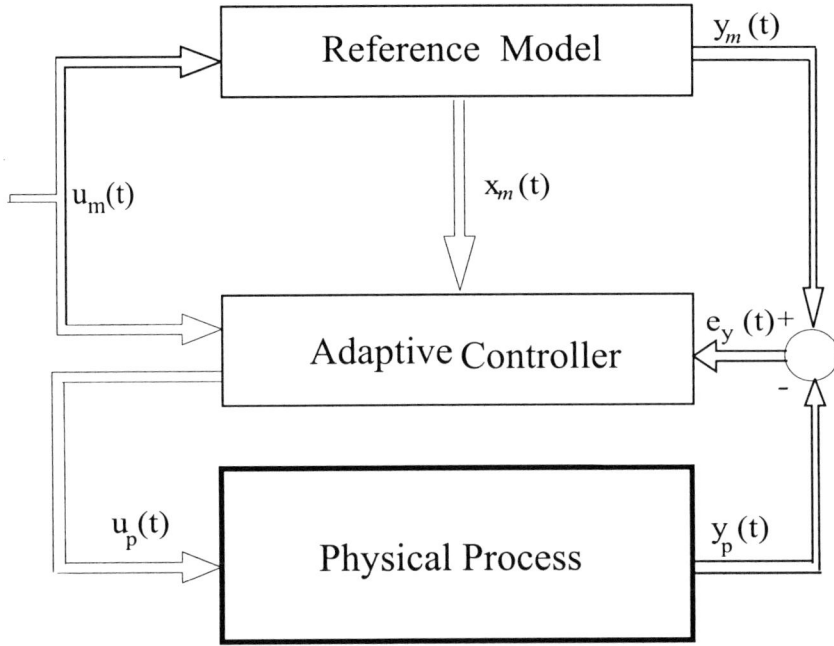

Figure 4.13: Basic principle of adaptive control.

However, within the welding community, the adaptive control scheme takes several forms, for example see Figure 4.14, which was proposed with the intention of developing an expert welding robot [528]. The overall scheme, self-explanatory, consists of various blocks THAT consist of several smaller blocks. Several other adaptive control schemes exist, especially for welding robots [580, 581, 576, 529]

An adaptive system with a focus on weld quality is given in [582]. Henderson, et al. [496, 494], reported a successful application of a pseudo-gradient adaptive algorithm for self-tuning a PI-based puddle width controller for a consumable electrode GMAW process. An adaptive control system for trajectory control (joint profile) was presented in [583]. An adaptive controller was developed where the controller gains were varied depending upon the nominal values of the current and WFS in [584].

A temperature control system was designed for a consumable elec-

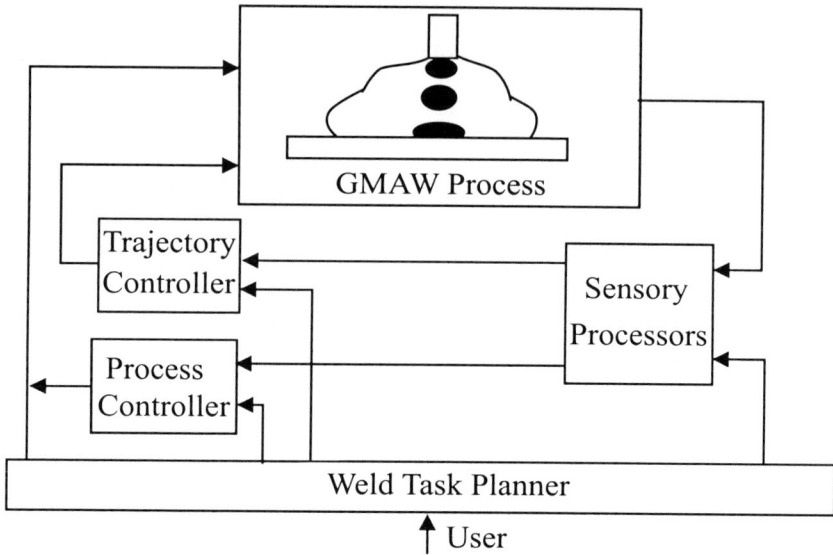

Figure 4.14: Alternative scheme of adaptive control.

trode GMAW process using an inexpensive optical system for measuring weld temperatures. The system used a pseudo gradient adaptive algorithm for self-tuning a PI bead temperature controller. The controller was designed and then illustrated with experimental data [585, 486]. A pseudo-gradient adaptive algorithm automatically tuned the gains on-line during the welding process. This method based on the work [586] is different from other adaptive control works [587, 588] in the sense that the pseudo-gradient algorithm does not depend upon the number of parameters or controller structure to the order of the plant.

An autoregressive moving average (ARMA) model relating torch travel rate and plate temperature was identified on-line using a MRAC system [487]. This MRAC scheme, shown in Figure 4.15, included a second-level feedback for controlling weld parameters, which indirectly determines weld bead geometry and other metallurgical properties of the workpiece.

In other investigations [589, 590, 587], the authors presented a method for dynamic modeling and control of two thermal character-

4.6. ADAPTIVE CONTROL

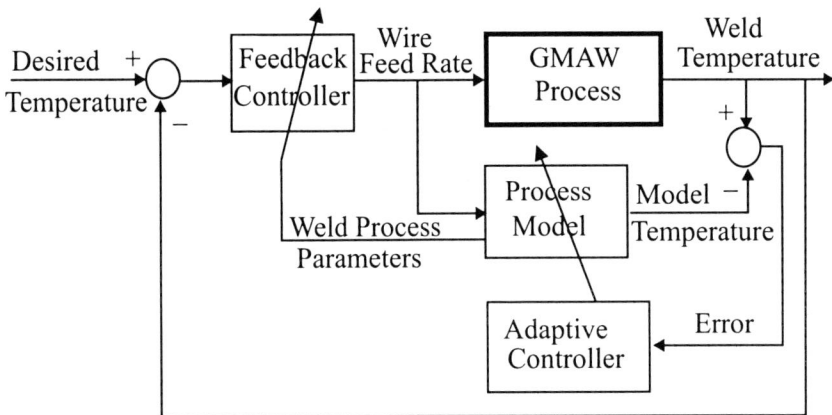

Figure 4.15: MRAC for plate temperature.

istics (heat affected zone and cooling rate) of a welding process. The basic model is a distributed parameter nonlinear process, but a lumped-parameter, locally linearized model is used to design a dead-beat adaptive control system. In [591], the authors developed a model for in-process control of thermally-activated material properties of weld. Also, see [587, 588] for the related work on multivariable adaptive control of thermal properties during welding. The work in [562] gives a similar treatment except the outputs to be controlled are bead width and depth.

In [592, 593, 594], Doumanidis devised an adaptive MIMO scheme to control both geometric and thermal characteristics of a weld based on lumped-parameter and distributed-parameter modeling and identification (see Figure 4.16 where Y_d is the desired output and Y_0 is the actual output.)

Further results on distributed parameter adaptive control are reported in [595] where an adaptive thermal control system using a Smith predictor to take care of the long transport delays of the thermal process [472] and in the present case the delay due to the temperature measurement at the torch location. Figure 4.17 shows the scheme, where T_d is the desired temperature distribution and Q is the heat input to weld process.

Doumanidis [593] gives an excellent account of some 25 works relat-

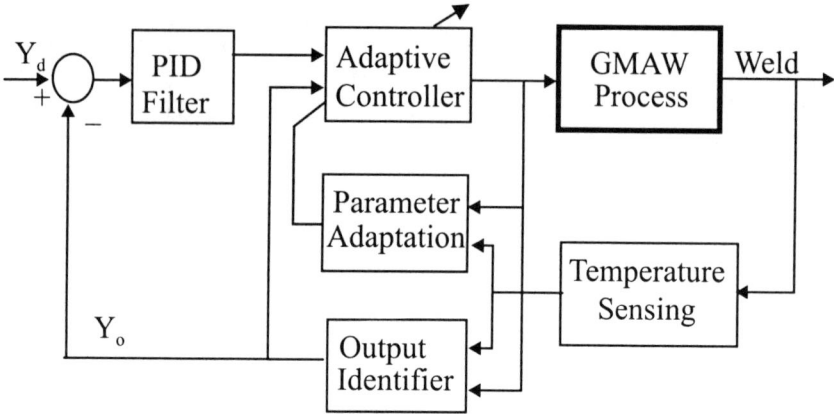

Figure 4.16: Adaptive thermal control system.

ing to control systems in welding research literature, classifying them in terms of the welding method, model, control technique, inputs, outputs, and sensors. Adaptation to changes in the geometry of the welding joint was carried out using a two-channel system for the control of movement of the robot [580, 581]. A system that maintains control over process parameters by continuously monitoring welding conditions through closed-loop feedback mechanisms was studied in [596] and the corresponding system is called adaptive or self-regulating system.

Other works in this area are [587, 588] where the authors found that the minimum number of adjustable parameters in the controller is dictated by the order of the plant in order to achieve not only stability but exact tracking of a reference signal.

Cook, et al. [469, 597] addressed the problem of adaptive and decoupling control of MIMO welding processes. The various input and output variables are not only related in a highly nonlinear fashion, but also they are also strongly coupled. Hence, there is a need for *decoupling* [598, 599] and *adaptive decoupling* [600] techniques for these MIMO welding systems, particularly for controlling direct weld parameters (DWP).

Kwak, in [601], considers the application of GMAW with deposition shape control for bead width control through the wire feed implemented in real time using Smith prediction to cope with sensor delays. Multi-

4.6. ADAPTIVE CONTROL

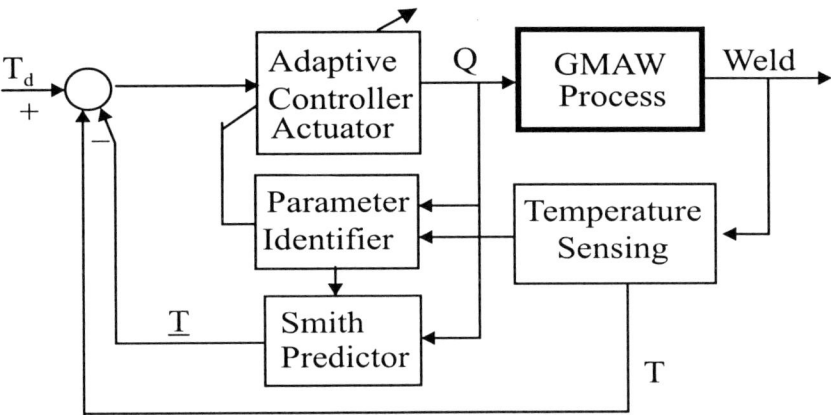

Figure 4.17: Adaptive thermal control system using a Smith predictor.

predictive adaptive control of arc welding trailing centerline temperature was performed in [602] for tor tackling the high level of uncertainty in the process.

4.6.1 ISU Adaptive Control Scheme

In order to achieve a desired mass (the transverse cross-sectional area of the deposited metal) and heat (given on a per unit length of weld) transfer values for a GMAW process, a fifth-order model described by highly nonlinear differential equations has been considered (ISU/INEEL [603]). After some simplification into a second-order model and linearization, we have a linear system with two inputs (open-circuit voltage and wire feed speed) and two outputs (arc current and arc voltage), where we assume the contact-tip-to-workpiece distance and weld torch speed to be constant. A direct model reference adaptive control (DMRAC) scheme based on the doctoral work of Ozcelik [604], was applied to the ISU/INEEL model of the GMAW process. With particular reference to our GMAW process, the DMRAC system is shown in Figure 4.18 [579]. This adaptive system was designed and implemented on the experimental facility at Idaho State University. This work is discussed in detail in Chapter 5 and in the reports [605, 579].

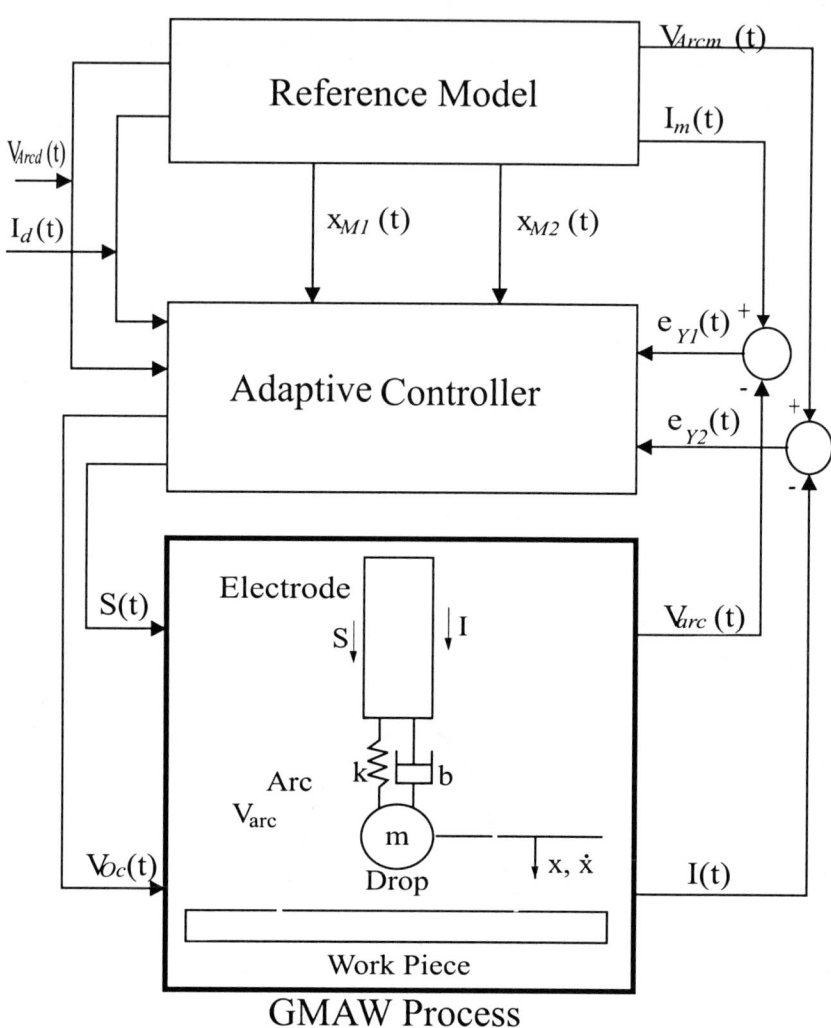

Figure 4.18: Direct model reference adaptive control of the GMAW process.

4.6. ADAPTIVE CONTROL

4.6.2 Other Works on Adaptive Control

An early development of an adaptive system for resistance arc welding control is presented in [606]. In [607, 608, 609], a systems approach to model the feedback and adaptive control of process variables in arc welding is discussed. A control scheme introduced by Cook [610] coordinates arc position sensing with manipulator motion to adaptively position the welding torch in response to unexpected changes in the weld joint trajectory.

A proof-of-concept of adaptive welding for flux cored arc and GTAW for small parts using feedforward control is demonstrated in [611]. Data processing problems associated with an adaptive control system for GMAW are discussed in [539, 612]. An adaptive control system for the GTAW process was designed in [613]. An awareness of the effects of weld joint dislocation on weld quality when utilizing robotics, where the joint dislocation is defined as the distance between the weld path of the robot and the centerline of the weld joint, is discussed in [614].

A robotic adaptive welding system is given in [615]. The development of a real-time adaptive spot welding control system was given [616]. The application of adaptive control theory to a GTAW process were demonstrated in [617, 618]. Two representative adaptive control schemes were used: the model reference adaptive control (MRAC) approach developed by Narendra and Lin [619] and the self-tuning adaptive control approach with pole placement [620]. An adaptive control system for weld penetration was designed in [621]. A good discussion of adaptive control of multivariable GMAW processes can be found in [561, 609]. The use of coaxial viewing for adaptive welding control of the GMAW process is developed in [622].

An adaptive control system for the GMAW process as a lab model is designed in [623]. The welding current and weld speed are controlled leading to high quality welds. Design of a model pool controller based on MRAC is given in [624]. In [625], an adaptive controller for multi-layer GMA welding of thick steel plates is presented.

A new paradigm for designing controllers for poorly-modeled systems with significant time-delay is introduced in [626, 627]. An adaptive, dead-beat compensator was developed, which is significantly different from the standard Smith predictor, and applied to the experiments on a GMAW testbed provided by the US Army Construction

Engineering Research Laboratory (CERL) in Champgaign, IL.

An adaptive control system for the GTAW process based on generalized predictive control was designed in [628, 629, 630, 631]. The controller predicts future outputs based on the present and future inputs [632]. The idea is applied to the GTAW process with non-minimum phase and variable large orders and delays in [633] and used in an adaptive scheme for robot welding in [634]. Other related works is found in [635, 628, 636, 633, 637].

An automatic welding system that can simultaneously control the bead height and back bead shape during one-sided metal active gas (MAG) welding with a backing plate was designed in [638] using a newly developed welding parameter control method in which only the wire feed-rate and welding voltage are adaptively controlled.

Tzafestas in [639] investigates the application of conventional and neural adaptive control schemes to GMAW with a review of four adaptive control techniques: MRAC, pseudo-gradient adaptive control (PAC), multivariable self-tuning adaptive control (STC), and neural adaptive control (NAC). In [640] an adaptive controller is introduced that is capable of "identifying" the arc sensitivity characteristic and adjusting the controller in real time for optimum response, without any a prior knowledge of the current/gain relationship.

4.7 Intelligent Control

In this section, we describe intelligent control of GMAW process. Intelligent control implies the use of neural networks, fuzzy logic, pattern recognition, expert systems, artificial intelligence (AI), and/or knowledge-based systems. These techniques, in particular neural networks and fuzzy logic, do not require precise mathematical modeling of the welding process, which is a stumbling block for all control techniques to be applied to welding processes.

4.7.1 Fuzzy Logic

Fuzzy logic is a concept based on set theory. Proposed by Zadeh [641], fuzzy logic was reengineered mostly by Japanese and then United States researchers. Fuzzy logic has been heavily used in control applications

4.7. INTELLIGENT CONTROL

[642]. Introducing a fuzzy set for, say molten pool width, one can define terms such "wide" and "narrow" and incorporate these terms into a fuzzy set with no boundary between "wide" and "narrow". See [624] for a molten metal pool controller based on fuzzy inference.

An experimental study of the application of fuzzy linguistic principles to control the peak surface temperature of the workpiece with wire feed rate as the input for an arc welding process is reported in [643]. Problems concerning the sensing of weld phenomena and the effects of power source characteristics on the stability of the arc are addressed in [644].

Sensing of the weld line using fuzzy control is presented in [645]. An application of fuzzy logic to spatial thermal control in fusion welding is reported in [646]. In this work, the theory of fuzzy sets was used as a general framework to interpret the uncertain arc signals and provide logic for control.

4.7.2 Neural Networks and Fuzzy Logic

Einerson, et al. [489], developed a control strategy for GMAW that employed an intelligent component in terms of a combination of an artificial neural network (ANN) for controlling electrode speed and torch speed and a fuzzy logic for controlling the reinforcement G and the input H (see Figure 4.8). In another multisensor-based control scheme [647], a neural network controller was developed as a bridge between the multiple sensor set and a conventional controller that provides independent control of the process variables such as torch speed, wire feed speed, CT, and open-circuit voltage.

In [648], the AI techniques involving ANNs and fuzzy logic were applied to address the problem of monitoring and controlling process variables such as welding power, torch velocity, and shielding gas to assure uniform and good quality welds in a GMAW process. In particular, the ANNs were applied to monitor weld pool geometry and the fuzzy logic controller was used to maintain arc stability and, hence, uniform weld quality. Also, in the experimentation, the fuzzy controller was found to be superior to the traditional PID controller.

Einerson, et al. [489], also developed a strategy for GMAW for controlling the reinforcement and weld bead centerline cooling rate, employing an intelligent component in terms of a combination of a neural network for controlling electrode speed and torch speed and a fuzzy logic controller for the reinforcement (G) and the input (H) (see Figure 4.8). Also, see other works by this group on intelligent sensing and control [647, 649, 650, 651].

Based on ANN and fuzzy logic, a self-learning neuro-fuzzy control system was developed for real-time control of pulsed GTAW in [652]. Here, an industrial TV camera was used as a sensor and by means of computer imaging techniques, the weldface width was estimated for use as a feedback signal. A block diagram employed by the authors is shown in Figure 4.19. Here, Y is the output, Y_d is the desired output,

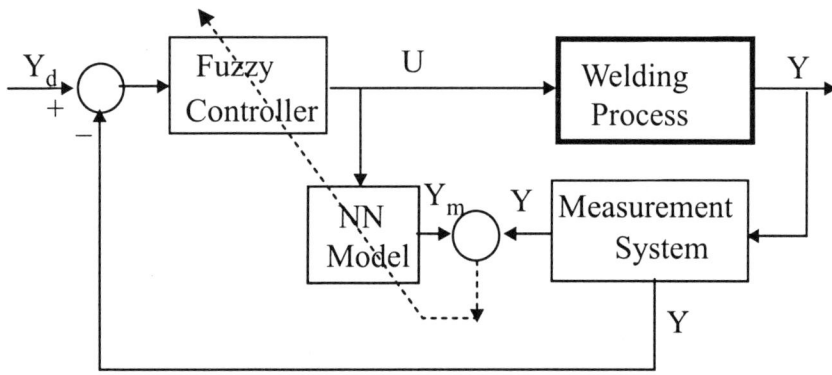

Figure 4.19: Self-learning fuzzy neural control system for arc welding processes.

Y_m is the model estimated by the neural network (NN), and U is the control input to the process.

Kovacevic and Zhang [653] used a feedback algorithm based on a neuro-fuzzy model for weld fusion to infer the back-side bead width from the pool geometry. A neuro-fuzzy model is one where the parameters of a fuzzy model are trained (adapted) by using neural networks [654]. In a typical experimental setup, the weld pool image is captured by a CCD camera and processed through an image processing unit, and then a neurofuzzy estimator provides the weld bead geometry (top-side and back-side widths), which is incorporated into a feedback algorithm to achieve the *desired* bead geometry, as shown in Figure 4.20.

Extensive results can be found on this and related topics by this

4.7. INTELLIGENT CONTROL

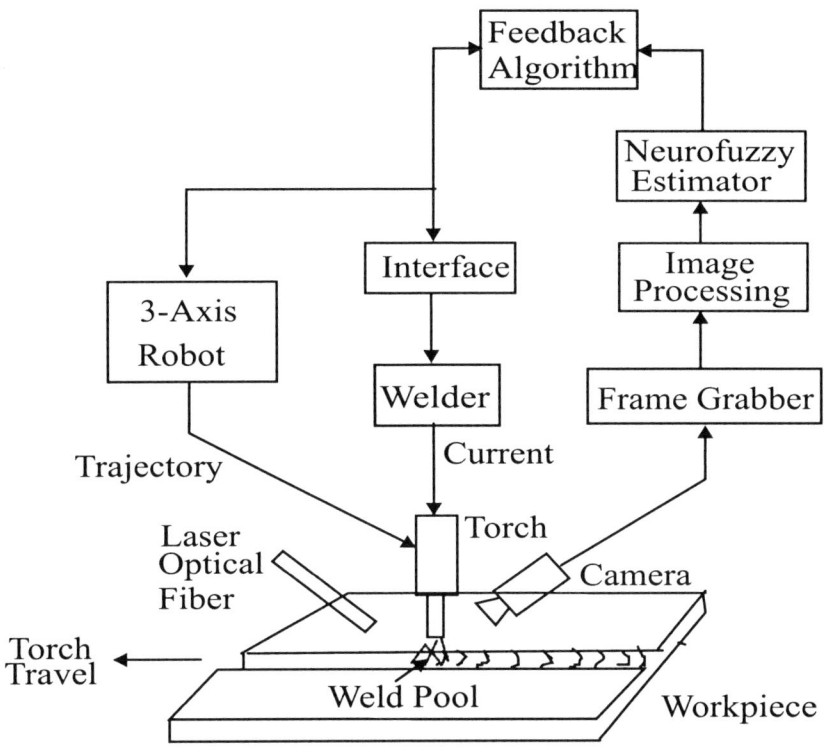

Figure 4.20: Experimental setup for neurofuzzy model-based control.

group in [655, 656, 657, 658, 633, 659, 660, 661]. Also, refer to [662] for the problem of tracking the welding line in an arm-type welding robot using fuzzy neural network.

On-line monitoring of weld defects for short-circuit GMAW based on the self-organizing feature map type of neural network was presented in [663]. It is based on the extraction of arc signal features as well as classification of the obtained features using SOM neural networks to get the weld quality information.

4.7.3 Knowledge-Based and/or Expert System

Expert systems in welding are intended to close the gap between a qualified operator who is inexperienced in welding and the skilled welder.

Incorporating extensive intuitive user interfaces and intelligent front-ends for encoding welding knowledge and expertise, robotic (adaptive) welding system using sensor feedback can compensate for part variations [664, 665, 666].

An AI system for automatic control of a narrow-gap GMAW process was developed in [667]. In order to eliminate the need for an experienced operator, the authors examined the application of AI based control. One such configuration is shown in Figure 4.21. In the ex-

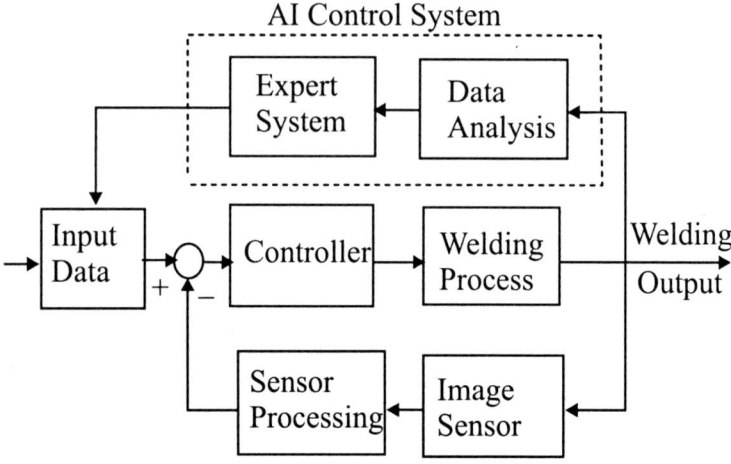

Figure 4.21: Artificial intelligence system for a welding process.

pert system the welding data is inferred both from the control codes and from the knowledge of the data base that is collected from the experience of skilled operators and from the measurement data.

An expert system called WELD-ASSIST for a robotic GMAW process used with low-carbon steel and mild steel is presented in [666]. The use of expert systems for adaptive control strategies were investigated in [668] for small batch arc welding to significantly improve the productivity, quality, and reliability of naval shipyard welding activities.

The development of an expert system for GMAW of aluminum and its alloys in Turbo-Pascal were developed in [669]. The expert system gives the complete procedure and provides recommendation on the type of power source, the type of welding current, the electrode angle, and a host of other welding parameters. In addition the neural networks

4.7. INTELLIGENT CONTROL

are used as target functions for genetic programming in order to find an optimized welding parameter set [670].

4.7.4 Other Works on Intelligent Control

In [671] pattern recognition techniques are employed for feedback position control using the arc signals and establishing a relationship between the electrode-to-workpiece spacing and arc signals. Intelligent control of welding processes in general can be found in [514, 672, 673].

An expert system-based control system for positional control system is presented in [674]. The Adaptweld System^{TM} is used in [538] to provide an expert welding system incorporating the knowledge of several skilled human welders in its information and control knowledge base. A hybrid hierarchical controller capable of compensating for incomplete modeling of the welding process was presented in [675] using a variety of expert systems, artificial neural networks and adaptive algorithms.

Intelligent sensing and control in arc welding processes in general and in GMAW in particular were discussed in [651]. The author draws an interesting distinction between the application of intelligent tools to process control examines various "objectives" such as process modeling, sensing, control theory, and artificial intelligence.

The application of neural networks to model a GTA welding process with experimental verification was reported in [676]. It was noted that such intelligent modeling techniques are needed for the application of Intelligent Processing of Materials (IPM) concepts to real manufacturing situations. A computer-aided-design (CAD)-based expert system for a GMAW process equipped with a six-axis industrial robot was presented in [677].

SmartWeld, a system for intelligent design and fabrication of welding components, developed by Sandia National Laboratories, Albuquerque, New Mexico was introduced in [678]. Siewert [679] discusses that welding problems such as the melting of the contact tube in GMAW can now be quickly corrected by employing automated intelligent control systems. Some other experimental results of intelligent/robust control and neural networks as applied to welding processes can be found in [680, 681, 682, 683, 684, 685].

Cook [686] considers three aspects of robotic arc welding of aluminum: coordinated motion control and weld path programming, penetration control, and joint tracking with through-the-arc sensing. A fuzzy logic system for process monitoring and quality evaluation in GMAW is designed in [687]. In this work, a fuzzy logic system that is able to recognize common disturbances during automatic GMAW using measured welding voltage and current signals is introduced.

4.8 Statistical Process Control and Quality Control

Here, we briefly review statistical process control, quality control, and quality assurance issues related to GMAW. Statistical quality control (SQC) is the application of statistical methods for the purpose of determining if a given component of production (input) is within acceptable statistical limits and if there is some result of production (output) that may be shown to be statistically acceptable to required specifications [688]. On the other hand, statistical process control (SPC) is the application of statistical methods for the purpose of determining if a given process is within the operating control parameters established by statistical procedures [689].

In [690, 691, 692], a statistical process control (SPC) technique was applied to a GMAW process to provide weld process quality control by using standard statistical process techniques, trending analysis, tolerance analysis, and sequential analysis [693]. Also refer to the work in [694] on SPC applied to GMAW.

A conceptual model of a pipeline welding quality control system was designed in [695, 696]. A weld process control system for computerized control and maintenance of the appropriate weld quality was designed in [697]. The information extracted from the real-time radiographic images about weld quality, supplemented by sensor data on weld current and voltage, was used for weld power-supply control. A discussion on the use of SPC for detecting defects in arc welding is given in [698].

A totally integrated weld quality monitoring system for GMAW was developed in [699] for recording, analyzing, and modifying welding parameters for quality verification of the weld and for tracing discontinuities. The application of SPC techniques to assess the quality of welds produced by the GMAW process is given in [700]. The control

4.9. OTHER CONTROL METHODOLOGIES AND ISSUES

charts for use in tracking voltage and current resulted in many false indications of acceptable welds and lack of indication of defective welds. Prediction of process parameters for GMAW by multiple regression analysis was reported in [701]. In this study a regression model was obtained from welding process parameters through the correlation of the parameters of the back-bead, to which an inverse transformation is performed.

On-line control of robotized GMAW was reported in [702]. The proposed control system aims to detect the most critical defects in industrial applications.

4.9 Other Control Methodologies and Issues

Under this section, we first review some specific control methodologies that have been applied to GMAW and then survey various issues not covered above.

4.9.1 Iterative Learning Control

Iterative learning control (ILC), a relatively a new technique within the arsenal of the control engineer, is a technique for improving the transient response and tracking performance of any physical system that is required to execute a particular operation repeatedly (such as a manipulator that might be programmed to do spot welding in an automobile manufacturing assembly line). By observing the error in the output response after each operation and using the error to modify the input signal to the system, ILC attempts to improve the system performance [703]. In other words, ILC is a technique for systems with repetitive or *iterative* operations, which are modified based on the observed error (or are programmed to *learn*) to control the input signal at each repetitive operation.

In another contribution to the development of ILC scheme to GMAW process [704], where the time interval between detachments of mass droplets from the end of an electrode is considered as a trial, the objective of ILC is to force the mass to detach at regular intervals with a uniform amount of mass in each detached droplet.

4.9.2 Feedback Linearization

Feedback linearization is a powerful techniques for analysis and design of nonlinear systems. The central idea of this approach is to algebraically transform the nonlinear system dynamics into a fully or partially linearized system so that the feedback control techniques could be applied [705, 706]. Note that this linearization technique, which is an exact state transformation and feedback, is entirely different from the conventional linearization based on Taylor series approximations.

In another investigation, the feedback linearization technique was applied to the fifth order, nonlinear model developed by ISU/INEEL researchers [603, 707, 708, 709] for a GMAW process. The fifth-order model (2.55) was first approximated into a second-order nonlinear model (2.61) in terms of two states current and stick out and three inputs: wire feed rate, open-circuit voltage and contact-tip-to-work piece distance. Using this second-order model and the steady-state models for heat and mass transfer [482], it was shown that it is possible to independently control current and arc length (effectively stick out) using the open-circuit voltage and wire feed speed. This result is described in more detail in the next chapter.

4.9.3 Relative Gain Array

In any study of control of multivariable physical systems, the process interaction is an important factor influencing the system behavior. A quantitative measure of interaction is needed to apply a multiloop controller and the relative gain array (RGA) is a useful technique for determining the appropriate loop pairing [710].

The RGA method was applied to the ISU/INEEL model [603, 579]. It was found that the correct pairing is that wire feed speed should be used to control the current and the open-circuit voltage should be used to control the arc voltage. Based on these loop pairings, several multi-loop controllers were designed. The complete details can be found in Chapter 5 and [711].

4.9.4 Other Works on Control

On-line control of arc welding processes by obtaining a relation in equation form between the inputs (such as arc applied voltage, current, wire

4.9. OTHER CONTROL METHODOLOGIES AND ISSUES 181

feed rate, welding gun position and speed) and outputs (such as weld bead dimensions) of the welding process was performed in [712]. Automatic precision TIG welding techniques were developed in [713] for welding all types of joints required on nuclear fuel elements by the Springfield Nuclear Power Development Laboratories (SNPDL), Salwich, Preston, UK.

Cook, in [714], gives a distributed microcomputer control system used in the programming, sensing, and feedback control of the welding process parameters. An interesting general discussion on the need for automation in welding processes aiming towards flexible manufacturing facilities is given in [715]. An automatic weld-line tracking system was developed in [716] by employing a light scanning technique using a laser and an image sensor for the sectional pattern of the joint groove.

A real-time machine vision-based feedback control system was designed in [717] to compensate for static and dynamic geometry variations as well as control of welding process parameters. A general-purpose, real-time seam tracking algorithm was developed in [529] for implementation on any six-degree-of-freedom robot, where the algorithm requires knowledge of only one point ahead to track a seam.

A process control system for arc welding applications was designed in [718] to provide advanced capabilities for tracking and analysis of welding variables. The requirements for second generation automatic welding systems capable of multipass welding, such as the machine intelligence capable of image perception with the ability to think strategically, were investigated in [719]. A process controller for vertical strip cladding using melt level sensing methods and a guide shoe design for the GMAW process were presented in [720].

An excellent tutorial and review type of discussion on modeling, sensing and control of welding processes were given in [469]. In [721, 722] it was pointed out that a simple automatic voltage control system, obtained by using gain scheduling technique, may be unstable over a wide range of current settings due to variations in arc sensitivity with current. A wire feed control system using a DC motor is given in [479].

The use of an infrared feedback signal for the automatic tracking of single V-groove prepared butt joints was discussed in [723]. In [722], a simple automatic voltage control (AVC) system that was unstable over a wide range of welding currents because of the arc sensitivity with current was designed. Digital feedback control of weld penetration of a GTAW using ANN for modeling was proposed in [724].

An automated robotic variable-polarity plasma arc welding (VP-PAW) for the Space Station Freedom Project (SSFP) is presented in [725]. The SSFP requires approximately 1.3 miles of aluminum welding for the final assembly. The VPPAW was chosen because of its ability to make defect-free welds in aluminum and the robotic VPPAW system was built by ABB Robotics, Inc., of Greenwood, SC, and installed at NASA's Marshall Space Flight Center.

A very good general presentation on the need for modeling and control of manufacturing processes in general and welding in particular is given in [726]. The author reviewed two decades of manufacturing control research in the ASME Journal of Dynamic Systems, Measurement and Control and found that there are only 25 articles published in the Journal on the whole of manufacturing, and out of them, there were only 6 papers on arc welding.

The work reported in [727] is an edited collection on sensors and control systems in arc welding, and, although a very good presentation of various topics, almost all the literature is limited to works in Japan. Control systems in general and welding process control in particular are discussed in [728] and [624], respectively.

A process control system for the arcing and short-circuiting phases and a study of its effect on spatter is presented in [729]. Future trends on control systems for arc welding processes, with an overwhelming response towards adaptive control, are discussed in [730]. Holm, in [731], develops a method for state space modeling of the whole manufacturing control system, including welding, and presents the use of state space models for improving processing control by articulating issues such as stability, disturbance compensation, hierarchical control, state estimators, and plant parameter estimation.

An arc welding penetration control system using quantitative feedback theory (QFT) is given in [732]. QFT is a unified theory that emphasizes the use of feedback for achieving desired robust system performance tolerances despite structured plant uncertainty and plant disturbances [733]. Other works in these areas can be found in the section on sensing, control and automation in [734]. A computer simulation of GMAW start-up was developed in [735] that accounts for the voltage-current characteristics of the welding arc, the welding power supply, and the interaction between the moving anode wire and the welding arc.

In [736], it was shown that an arc discharge in a GMAW welding process with fusible electrodes starts before a short-circuit bridge, made by a metal drop between the electrodes, is broken. The possibility of this premature ignition is proved by voltage and current measurements, by analysis of the electrical field near the neck of the drop, and also by simulation of the increase in current in an electrical discharge.

A unique excitation, sensing, and control system was developed in [737] to predict and control the state of penetration during the GTAW process. Excitation of the molten pool is accomplished by synchronously modulating the arc force in phase with the weld pool's own natural frequency using a phase-locked loop (PLL) technique. Regulation of GMA welding thermal characteristics via a hierarchical MIMO predictive control scheme that assures stability has been presented in [738]. The work proposes a hierarchical predictive control scheme for the metallurgical characteristics of GMAW.

Numerical analysis of the dynamics of droplet growth in GMAW was studied by Zhang in [739]. Feedback of droplet transfer is pursued as a solution to produce sound GMAW welds. Zhang has also developed a robust control algorithm to control the pulsed gas metal arc welding process [740]. In a recent study [741], Zhang also proposed a modified active control to ensure a specific type of desirable repeatable metal transfer modes.

Other results that have been reported at various experimental facilities for implementing automatic controllers for the GMAW process include [526, 501], [478], [489], [603, 579], [742, 561, 560], [633, 637], [626, 743, 744, 745, 627].

4.10 Safety and Environmental Issues

Some of the potential safety hazards associated with welding operations are radiation, visible, infrared and ultraviolet light; ionizing radiation (x-rays) (due to EBW); toxic gases (due to arc processes) and noise (due to friction and plasma). Hence, the welding operator is normally protected by means of proper protective clothing, head gear with eye protection, local screening and ventilation. A number of different hazards present in electric arc welding and their effective control is achieved by conventional environmental engineering solutions. Control of welding fumes at the source rather than by local exhaust ventilation is presented in [746]. In particular, the effect of the rate of fume generation of the various welding parameters such as voltage, current, wire feed rate, shielding gas and the material to be welded, were studied.

4.11 Classification of References by Section

Here, we provide a table containing the various references according to each section of this chapter. This will provide a ready reference to the interested reader to search for relevant references in each section.

4.11. CLASSIFICATION OF REFERENCES BY SECTION

Table 4.1: Section by Section List of References

Section	Reference Numbers
4.0 Gas Metal Arc Welding	[468]
4.1 Manual Control Techniques	
4.2 Automation or Automatic Welding	
4.3 Automatic or Feedback Control Techniques	[469]-[475]
4.4 Control of Process Variables	[476]-[549]
4.5 Classical Control: PI, PID, and Others	[482],[488], [550]-[559]
4.6 Multivariable Control	[560]-[575]
4.7 Optimization and Optimal Control	[471],[494],[547], [576]-[579]
4.8 Adaptive Control	[469],[472],[486], [487],[494],[496] [528],[529],[539], [561],[562],[577] [580]-[641]
4.9 Intelligent Control	[489],[514],[538], [625],[634] [642]-[688]
4.10 Statistical Process Control and Quality Control	[689]-[703]
4.11 Other Control Applications and Issues	[469],[475],[482], [529],[580], [604],[625], [704]-[742]
4.12 Safety and Environmental Issues	[743]
4.13 Experimental Facilities	[479],[489],[501], [526],[560],[561] [580],[604],[627], [628],[634],[638] [744]-[746]

References List for Chapter 4

[468] J. Norrish. Advanced welding processes. Institute of Physics Publishing. Bristol, UK, 1992.

[469] G. C. Cook, K. Anderson and R. J. Barrett. Feedback and adaptive control in welding. Proceedings of the 2nd International Conference on Trends in Welding Research. Gatlinburg, TN. May. S. A. David and J. M. Vitek. Key Note Address. Pages 891-903, 1989.

[470] C. L. Phillips and R. D. Harbor. Feedback Control Systems, Third Edition. Prentice Hall. Englewood Cliffs, NJ, 1996.

[471] F. L. Lewis and V.L. Syrmos. Optimal Control, Second Edition. John Wiley & Sons. New York, NY., 1995.

[472] K. J. Astrom and B. Wittenmark. Adaptive Control. Addison-Wesley Publishing Company, Inc. Reading, MA. Second, 1995.

[473] M. J. Grimble. Robust Industrial Control: Optimal Design Approach for Polynomial Systems. Prentice Hall. Englewood Cliffs, NJ., 1994.

[474] M. M. Gupta and N. K. Sinha. Intelligent Control Systems: Theory and Applications. IEEE Press. New York, NY., 1996.

[475] R. L. O'Brien. Welding handBook: Welding processes. Volume 2, Edition Eighth. Miami, FL. American Welding Society, 1991.

[476] K. S. Ogilvie. Modeling of the gas metal arc welding process for control of arc length. University of Michigan. Ann Arbor, MI. PhD Dissertation, 1991.

[477] C. A. Johnson and A. M. Sciaky. System for controlling length of welding arc. US Patent 3236997, 1966.

[478] R. B. Madigan. Control of gas metal arc welding using arc light sensing. School Colorado School of Mines. Golden, CO., 1994.

[479] B. W. Greene. Arc current control of a robotic welding system: modeling and control system design. Institution Illinois Univ. at Urbana-Champaign, Coordinated science Lab. Champaign, IL. DC-114-UILU-ENG-89-2227, 1990.

[480] G. Linden and G. Lindskog and L. Nilsson. A control system using optical sensing for metal-inert gas arc welding. Proceedings of an International Conference on Developments in Mechanized, Automated and Robotic Welding. London, UK. November. Pages P17-1-P17-6, 1980.

[481] G. J. Ogilvie and I. M. Ogilvy. The pulsed GMA process in automatic welding. Proceedings of 31st Annual Conference of The Australian Welding Institute. Sydney, Australia. Pages 16-19. October, 1983.

[482] H. B. Smartt and C. J. Einerson. A model for heat and mass input control in gas metal arc welding. Welding Journal. Volume 72(5), Pages 217s–229s. May, 1993.

[483] A. Ditschun and B. Zajaczkowski and D. Dorling. Pulsed FM-GMA welding. Proceedings of the Conference on Welding for Challenging Environments. Pages 21-29. Toronto, Canada. Oct. 15-17, 1985.

[484] A. Ditschun and D. Dorling and A. G. Glover and B. A. Graville and B. Zajaczkowski. The development and application of pulsed FM-GMA welding. Proceedings of EWI 1st International Conference on Advanced Welding system. P. T. Houldcroft. Pages 321-329. London, UK. November, 1985.

REFERENCES LIST FOR CHAPTER 4

[485] W. G. Essers and M. R. M. Van Gompel. Arc control with pulsed GMA welding, Welding Journal. Pages 26-32, June, 1984.

[486] D. V. Nishar, J. L. Schiano and W. R. Perkins and R. A. Weber. Adaptive control of temperature in arc welding. IEEE Control Systems Magazine. Volume 14(4), Pages 4-12, August, 1994.

[487] D. A. Dornfeld and M. Tomizuka and G. Langeri. Modeling and adaptive control of arc welding processes. Measurement, Control in Batch Manufacturing. ASME. D. E. Hardt. New York, NY. Pages 53-64. November, 1982.

[488] D. F. Farson and R. W. Richardson and R. J. Mayham. Numerical simulation of feedback control of arc welding processes. Proc. of the Intl. Conf. on Trends in Welding Research. Gatlinburg, TN. May, 1986.

[489] C. J. Einerson and H. B. Smartt and J. A. Johnson and P. L. Taylor and K. L. Moore. Development of an intelligent system for cooling rate and fill control in GMAW. Proceedings of the ASM 3rd International Conference on Trend in Welding Research. Gatlinburg, TN. S. A. David and J. M. Vitek. Pages 853-857. June, 1992.

[490] G. E. Cook. Intrinsic thermocouple monitors welding. Metals Progress. Volume 93. Pages 176-180, 1968.

[491] B. E. Bates and D. E. Hardt. A real-time calibrated thermal model for closed-loop weld bead geometry control. ASME Journal of Dynamic Systems, Measurement and Control. Volume 107, Pages 25-33. March, 1985.

[492] A. J. Aendenroomer. Weld Pool Oscillation for Penetration Sensing and Control. School Technical School of Delft University. Delft, The Netherlands, 1996.

[493] LabVIEW User Manual for Windows. National Instruments Corporation. Austin, TX, 1994.

[494] J. L.Schiano, J. H. Ross and R. A.Weber. Modeling and control of puddle geometry in gas metal arc welding. Proceedings of the

1991 American Control Conference. Boston,MA. Pages 1044-1049, 1991.

[495] D. E. Henderson. Adaptive control of an arc welding process. Institution University of Illinois. Urbana, IL. Rep. DC-148, UILU-ENG-90-2220, 1990.

[496] D. E. Henderson, P. V. Kokotovic, J. L. Schiano and D. S. Rhode. Adaptive control of an arc welding process. Proceedings of the American Control Conference. Pages 655-660. Boston, MA. June, 1991.

[497] D. E. Henderson, P. V. Kokotovic, J. L. Schiano and D. S. Rhode. Adaptive control of an arc welding process. IEEE Control Systems Magazine. Volume 13. Pages 49-53. February, 1993.

[498] R. S. Baheti, K. B. Haefner and L. M. Sweet. Operational performance of vision-based arc welding Robot control systems. Institution General Electric Company. Schenectady, NY. November. Pages 1-8, 1984.

[499] R.S. Baheti. Vision processing and control of robotic arc welding system. Proceedings of the 24th IEEE Conference on Decision and Control. Pages 1022-1024. Ft. Lauderdale, FL. December, 1985.

[500] M. B. Hale and D. E. Hardt. Multi-output process dynamics of GMAW: limits to control. Proceedings of the ASM 3rd International Conference on Trends in Welding Research. Gatlinburg, TN. Pages 1015-1020. June, 1992.

[501] P. Banerjee, S. Govardhan, H. C. Wikle, J. Y. Liu and B. A. Chin. Infrared sensing for on-line weld geometry monitoring and control. ASME Transactions, Journal of Engineering for Industry. Volume 117, Pages 323-330. August, 1995.

[502] A. R. Vorman and H. Brandt. Feedback control of GTA welding using puddle width measurement. Welding Journal. Volume 55(9), Pages 742-749, 1976.

REFERENCES LIST FOR CHAPTER 4

[503] K. Masubuchi, D. E. Hardt, H. M. Paynter and W. Unkel. Improvement of reliability of welding by in-process sensing and control. Proceedings of Trends in Welding Research in The United State. New Orleans, LA. Pages 667-688, November16-18, 1981.

[504] R. W. Richardson, A. Gutow, R. A. Anderson and D. F. Farson. Coaxial arc weld pool viewing for process monitoring and control. Welding Journal. Pages 43-50. March, 1984.

[505] W. F. Clocksin, J. S. E. Bromley, P. G. Davey, A. R. Vidler and C. G. Morgan. An implementation of model-based visual feed-back for robot arc welding of thin sheet steel. International Journal of Robotics Research. Volume 4(1), Pages 13-26, 1985.

[506] D. Stone, J. S. Smith and J. Lucas. Sensor for automated weld bead penetration control. Measurement Science and Technology. Volume 1, Pages 1143–1148, 1990.

[507] J. A. Johnson, N. M. Carlson, H. B. Smartt and D. E. Clark. Process control of GMAW: Sensing of metal transfer Mode. Welding Journal. Volume 70, Pages 91s-99s. April, 1991.

[508] C. L. Phillips and H. T. Nagle. Digital Control System Analysis and Design, Third Edition. Prentice Hall. Englewood Cliffs, NJ., 1995.

[509] J. S. Bendant and A. G. Piersol. Random Data: Analysis and Measurement Procedures. Wiley-Interscience. New York, NY., 1971.

[510] A. P. Bennett and C. J. Smith. Improving the consistency of weld penetration by feedback control. Proceedings of Fabrication and Reliability Welded Process Plant. The welding Institute, London, UK. Pages 13-19, 1976.

[511] P. Boughton, G. Rider and C. J. Smith. Feedback control of weld penetration. Proceedings of 4th International Conference on Advances in Welding Processes. Harrogate, England. Pages 203-215. May, 1978.

[512] H. Nomura, T. Yohida and K. Tohno. Control of weld penetration. Metal Construction. Volume 8, Pages 244-246. June, 1976.

[513] D. E. Hardt, D. A.Garlow and J. B. Weinert. A model of full penetration arc welding for control system design. Proceedings of the Winter Annual Meeting of the ASME on Control of Manufacturing Processes and Robotic Systems. D. E. Heart and W. J. Book. ASME, New York. Pages 121-135. November, 1983.

[514] W. M. McCampbell, G. E. Cook, L. E. Nordholt and G. J. Merrick. The development of a weld intelligence system. Welding Journal. Volume 45(3), Pages 139s-144s. March, 1966.

[515] N. H. Madsen and B. A. Chin. Automatic welding: Infrared sensors for process control. Proceedings of 11th Conference on Production Research and Technology. Pittsburgh. Organization PA, Society of Manufacturing Engineers. Pages 277-285. May, 1984.

[516] J. L. Fihey, P. Cielo and G. Begin. On-line weld penetration measurement using an infrareds sensor. Proceedings of International Conference Welding in Energy-Related Projects. Toronto, Canada. Pages 177-188. September, 1983.

[517] G. Rider. On line measurement of weld pool surface size. Proceedings of Welding and Fabrication in the Nuclear Industry. London, UK. Organization British Nuclear Energy Society. Pages 351-359, April, 1979.

[518] G. Rider. Control of weld pool size and position for automatic and robotic welding. Proceedings of the SPIE Third International Conference on Robot Vision and Sensory Controls RoViseC3. Cambridge, MA. Volume 449, Pages 381-389. November, 1983.

[519] R. A. Richardson, D. A. Gutow and S. H. Rao. A vision based system for arc weld pool size control. Measurement and Control for Batch Manufacturing. D. E. Hardt. ASME. New York, NY. Pages 65-75. November, 1982.

REFERENCES LIST FOR CHAPTER 4

[520] R. R. Stroud and R. Fenn. Microcomputer control of weld penetration. Proceedings of 31st Annual Conference on Welding and Computers. Sydney, Australia. Pages 10–14. October, 1983.

[521] J. M. Katz and D. E. Hardt. Ultrasonic measurement of weld penetration. Proceedings of Control of Manufacturing Processes and Robot Systems. D. E. Hardt and W. J. Book. Organization ASME. New York. Pages 79-95, November, 1983.

[522] K. Andersen, G. E. Cook, R. J. Barnett and A. M. Strauss. Synchronous weld pool oscillation for monitoring and control. IEEE Transactions on Industry Applications. Volume 33(2), Pages 464-471. March/April, 1997.

[523] R. J. Barnett, G. E. Cook, J. D. Brooks and A. M. Strauss. A weld penetration control system using synchronized current pulses. Proceedings of the ASM 4th International Conference on Trends in Welding Research. Gatlinburg, TN. H. B. Smartt, J. A. Johnson and S. A. David. Pages 727-732. June 5-8, 1995.

[524] E. A. Gladkov and I. A. Guslistov. The dependence of the intensity of the radiant flux on the parameters of the weld pool in systems for the automatic control of penetration. Automatic Welding. Volume 30. Pages 5-8, 1977.

[525] E. H. Daggett. Feedback control to improve penetration of dip transfer MIG. Proceedings of the EWI 1st International Conference on Advanced Welding Systems. P. T. Houldcroft. Pages 282-285. London, UK. November, 1985.

[526] W. H. Chen and B. A. Chin. Monitoring joint penetration using infrared sensing techniques. Welding Journal. Volume 69(4), Pages 181s-185s, 1990.

[527] M. Rioux. Laser range finder based on synchronized scanners. Applied Optics. Volume 23(21), 1984.

[528] P. Sicard and M.D. Levine. An approach to an expert robot welding system. IEEE Transactions on Systems, Man and Cybernetics. Volume 18(2), Pages 204-222. March/April, 1988.

[529] P. K. Khosla, C. P. Neuman and M. Prinz. An algorithm for seam tracking applications. The International Journal of Robotics Research. Volume 4(1), Pages 27-41, 1985.

[530] D. C. Verdon, D. Langley and M. H. Moscardi. Adaptive welding control using video signal processing. Proceedings of the International Conference on Developments in Mechanized, Automated and Robot Welding. London, UK. November, Pages P291-P2911, 1980.

[531] R. W. Richardson and R. A. Anderson. Weld butt joint tracking with a coaxial viewer based weld vision system. Proceedings of ASME Winer Annual Meeting on Control of Manufacturing Process and Robotic Systems. ASME. D. E. Hardt and W. J. Book. Boston, MA. Pages 107-119, November, 1983.

[532] R. W. Richardson and C. C. Conrardy. Coaxial vision-based control of GMAW. Proceedings of ASM 3rd International Conference on Trends in Welding Research. Gatlinburg, TN. S. A. David and J. M. Vitek. Pages 957-961. June, 1992

[533] H. Fujimura, E. Ide and H. Inoue. Joint tracking control sensor of GMAW: Development of method and equipment for position sensing in welding with electric arc signals (Report 1). Transactions of the Japan welding Society. Volume 18(1), Pages 32-40. April, 1987.

[534] H. Fujimura, E. Ide and H. Inoue. Weave amplitude control sensor of GMAW: Development of method and equipment for position sensing in welding with electric arc signals (Report 2). Transactions of the Japan Welding Society. Volume 18(1), Pages 41-45. April, 1987.

[535] M. Tomizuka. Design of digital tracking controller for manufacturing applications. Proceedings of the Winter Annual Meeting of the ASME on Control Methods for the Manufacturing Processes. Chicago, IL. D. E. Hardt. Nov.-Dec, 1988 Pages 71-78.

[536] T. Araya and S. Saikawa. Recent activities on sensing and adaptive control of arc welding. Proceedings of the 3rd International

REFERENCES LIST FOR CHAPTER 4   195

Conference on Trends in Welding Research. S. A. David and J. M. Vitek. Gatlinburg, TN. Pages 833-842, June, 1992.

[537] R. W. Gellie. Sensing for automated welding. Proceedings of the 31st Annual Conference on Welding and Computers. Sydney, Australia. Pages 193-200. October, 1983.

[538] W. J. Kerth Jr. Knowledge-based expert welding. Proceedings of Robots9 Conference. Pages 5-98-5-110. June, 1985.

[539] G. Nachev, B. Petkov and L. Blagoev. Data processing problems for gas metal arc GMA welder. Proceedings of the SPIE 3rd International Conference on Robotic Vision and Sensory Controls RoViSEC3. Volume 449. Cambridge, MA. D. P. Cassssent and E. L. Hall. Pages 291-296. November, 1983.

[540] Y. Arata and K. Inoue. Automatic control of arc welding by monitoring the molten pool. Transactions of Japan Welding Research Institute (JWRI) . Volume 1(1), Pages 99s-113s, 1972.

[541] Y. Arata and K. Inoue. Automated control of arc welding (report II)-optical sensing of joint configuration. Transaction. Jap. Welding Res. Institute (JWRI). Volume 2(1), Pages 87s-101s, 1973.

[542] Y. Arata and K. Inoue. Automatic control arc welding (report IV). Transactions Japan Welding Research Institute (JWRI). Volume 4(2), Pages 101s-104s, 1975.

[543] Y.M. Zhang, E. Liguo, and R. Kovacevic. Active metal transfer control by monitoring excited droplet oscillation. Welding Journal (Miami, Fla). Volume77(9), Pages388s-395s. September, 1998.

[544] H. Mecke, W. Fischer, I. Merfert, S. Nowak, U. Dilthey, U. Reisgen, L. Stein and H. Bachem. Combined primary and secondary-clocked, computer-controlled welding power supplies with higher dynamic ratio. Schweissen and Schneiden Welding & Cutting.. Volume50(4), PagesE67-E69. April, 1998.

[545] J. Tao and P. Levick. Assessment of feedback variables for through the arc seam tracking in robotic gas metal arc welding .

Proceedings of Feedback Variable for through the arc seam tracking in robotic gas metal arc welding. Piscataway, New Jersey. Pages 3050-3052, 1999.

[546] D. Barborak, C. Conrardy, B. Madignan and T. Paskell. Through-arc process monitoring techniques for control of automated gas metal arc welding. Proceedings - IEEE International Conference on Robotics and Automation. Piscataway, New Jersey. Pages 3053-3058, 1999.

[547] S. Subramaniam, D.R. White, J.E. Jones, and D.W. Lyons. Experimental approach to selection of pulsing parameters in pulsed GMAW. Welding Journal (Miami, Fla). Volume78(5), Pages166s-172s, 1999.

[548] Andrew Cullison. Get that Spatter under Control. Welding Journal. Volume78(4), Pages43-45, 1999.

[549] M. J. Hermans and G. O. Den. Process behavior and stability in short circuit gas metal arc welding. Welding Journal. Volume78(4), Pages137s-141s, 1999.

[550] M. Amin and N.-Ahmed. Synergic control in MIG welding 1 - parametric relationships for steady DC open arc and short circuiting arc operation. Metal Construction. Volume 19, Pages 22-28, 1987.

[551] M. Amin and N.-Ahmed. Synergic control in MIG welding 2 - Power-current controllers for steady state DC open arc operation. Metal Construction. Volume 19(6), Pages 331-340. June, 1987.

[552] H. B. Smartt, C. J. Einerson, A. D. Watkins and R. A. Morris. Gas metal arc process sensing and control. Proceedings of an International Conference on Trends in Welding Research. Gatlinburg, TN. ASME. Pages 461–465, 1986.

[553] H. B. Smartt, C. J. Einerson and A. D. Watkins. Computer control of gas metal arc welding. Computer technology in Welding, Cambridge, UK. Cambridge, UK. June, 1988.

REFERENCES LIST FOR CHAPTER 4

[554] H. B. Smartt, C. J. Einerson and A. D.watkins. Methods for controlling gas metal arc welding. EG&G Idaho, Inc., Idaho falls, ID, howpublished Patent, 1989.

[555] H. B. Smartt, A. D. Watkins and M. D. Light. Sensing and control problems in gas metal welding. Proceedings of the ASM 2nd International Conference on Trends in Welding Research. Gatlinburg, TN. May. S. A. David and J. M. Vitek. Pages 917-921, 1989.

[556] A. D. Watkins. Heat transfer efficiency in gas metal arc welding. University of Idaho. Moscow, ID, 1989.

[557] G. E. Cook and P. C. Levick. Microcomputer control of joint tracking system for arc welding. Proceedings of the Winter Annual Meeting of ASME on Computer Applications in Manufacturing Systems. Chicago, IL. Editor W.R. De Vries. Pages 93=101, November, 1980.

[558] G. Linden and G. Lindskog. A control system using optical sensing for metal-inert gas arc welding. Proceedings of TWI Conference. November, 1980.

[559] E. Kannatey-Asibu, Jr. Analysis of the GMAW process for microprocessor control of arc length. Transactions of the ASME, Journal of Engineering for Industry. Volume 109, Pages 172-176. May, 1987.

[560] D. E. Hardt. Welding process modeling and re-design for control. Proceedings of 4th US/Japan Symposium on Flexible Automation. San Francisco, CA. July. Pages 275-281. 1992.

[561] D. E. Hardt. Modeling and control of welding processes. Proceedings of the Fifth Conference on Modeling of Casting and Welding Processes. Davos, Switzerland. The Minerals, Metals and Materials Society. M. Rappaz and M. Ozgu and K. W. Mahin. Pages 287-303, 1990.

[562] J. B. Song and D. E. Hardt. Simultaneous control of bead width and depth geometry in gas-metal arc welding. Proceedings

of Third International Conference on Welding Research, ASME. Gatlinburg, TN. June, 1992.

[563] R. Masmoudi. Process decoupling for simultaneous control of heat affected zone and width in GMAW. Department of Mechanical Engineering, MIT. September, 1992.

[564] R. Masmoudi and D. E. Hardt. Multivariable control of geometric and thermal properties in GTAW. Proceedings of 3rd ASM Conference Trends welding Res.. Gatlinburg,TN. June, 1992.

[565] R. Masmoudi and D. E. Hardt. High-frequency torch weaving for enhanced welding controllability. Proceedings of the ASM 3rd International Conference on Trends in Welding Research. S. A. David and J. M. Vitek. Gatlinburg, TN. June. Pages 931-936, 1992.

[566] J. B. Song and D. E. Hardt. Dynamic modeling and adaptive control of the gas metal arc welding process. Transactions of the ASME, Journal of Dynamic Systems. Measurement and Control. Volume 116(3), Pages 405-413. September, 1994.

[567] J. B. Song and D. E. Hardt. Development of a heat-transfer based depth estimator for real-time welding control. Proceedings of ASME, Symposium on Manufacturing Process Modeling and Control. November, 1990.

[568] J. B. Song and D. E. Hardt. Multivariable adaptive control of bead geometry in GMA welding. Proceedings of ASME Winter Annual Meeting of the American Society of Mechanical Engineers. Atlanta, GA. December, 1991.

[569] J. B. Song and D. E. Hardt. A thermally based weld pool depth estimator for real-time control. Proceedings of 3rd International Conference on Trends in welding research. Gatlinburg, TN. Pages 975–980. June, 1992.

[570] J. B. Song and D. E. Hardt. Application of adaptive control to arc welding process. Proceedings of the American Control Conference. San Francisco, CA. Pages 1751–1755, June, 1993.

REFERENCES LIST FOR CHAPTER 4

[571] J. P. Huissoon, H. W. Kerr, S. Bedi, D. C. Wechman and W. P. Stefanuk. Design of an integrated robotic welding system. Proceedings of the 3rd International Conference on Trends in Welding Research. S. A. David and J. M. Vitek. Gatlinburg, TN. Pages 915-926, June, 1992.

[572] J. P. Huissoon, D. L. Strauss, J. N. Rempel, S. Bedi and H. W. kerr. Multi-variable control of robotic gas metal arc welding. Journal of Materials Processing Technology. Volume 43(1), Pages 1-12. June, 1994.

[573] W. L. Brogan. Modern Control Theory. Prentice Hall. Englewood Cliffs, NJ. Third Edition, 1991.

[574] W. P. Stefanuk, J. N. Rempel, J. P. Huissoon and H. W. Kerr. A multivariable approach to the control of fillet weld dimensions. Proceedings of the ASM 3rd International Conference on Trends in Welding Research. Gatlinburg, TN. June. S. A. David and J. M. Vitek. Pages 889-893, 1992.

[575] L. Ljung. System Identification: Theory for the User. Prentice Hall. Englewood Cliffs, NJ., 1987.

[576] M. Tomizuka, D. Dornfeld and M. Purcell. Application of microcomputers to automatic weld quality control. Transactions of the ASME: Journal of Dynamic Systems, Measurement and Control. Pages 62–68, Volume 102(2). June, 1980.

[577] S. Ray, M. Bhattacharyya and S. N. Banerjee. Investigation into optimal welding variables for arc welded butt joints. Welding Journal. Pages 39s-44s. February, 1986.

[578] K. Kondoh and T. Ohji. Algorithm based on convex programming method for optimum heat input control in arc welding. materials Transactions. Volume39(3), Pages420-426. March, 1998.

[579] K.L. Moore, D.S. Naidu, S. Ozcelik, R. Yender and J. Tyler. Advanced Welding Control Project: Annual Report, FY97. Measurement and Control Engineering Research Center. Idaho State University, Pocatello, ID. July, 1997.

[580] K. B. Alekseev, V. A. Afonin and M. M. Fishkis. A two-channel control system for an adaptive welding robot. Automatic Welding(GB). Volume 32(12), Pages 25-28. December, 1979.

[581] K. B. Alekseev, M. M. Fishkis and V. V. Fokin. The adaptive control of a welding robot. Welding Production(GB). Volume 27(9), Pages 4-8. September, 1980.

[582] P. Boughton, G. Rider, and G. J. Smith. Towards the automation of arc welding. CEGB Research, Central Electricity Generating Board. London, UK, June, 1979.

[583] P. sicard and M. D. Levine. Automatic joint recognition and tracking for robotic arc welding. Proceedings of IEEE Montech '87-Compint'87. Montreal, Canada. Pages 290–293. November, 1987.

[584] T. P. Quinn and R. B. Madigan. Adaptive arc length controller design for GMAW. Proceedings of American Welding Society International Conference on Modeling and Control of Joining Processes. 1993.

[585] D. V. Nishar. Feedback control of bead temperature in gas metal arc welding. University of Illinois. Urbana, IL., 1992.

[586] D. Rhode and P. V. Kokotovic. Parameter convergence conditions independent of plant order. Proc. of 1989 American Control Conference. Pittsburgh, PA. Pages 981-986, 1989.

[587] C. C. Doumanidis and D. E. Hardt. Simultaneous in-process control of heat-affected zone and cooling rate during arc welding. Welding Journal. Volume 69, Pages 186s-196s. May, 1990.

[588] C. C. Doumanidis and D. E. Hardt. Multivariable adaptive control of thermal properties during welding. Transactions of the ASME, Journal of Dynamic Systems Measurement and control. Pages 82-92. Volume 113(1). March, 1991.

[589] C. C. Doumanidis. Modeling and Control of Thermal Phenomena in Welding. MIT. Cambridge, MA, 1988.

REFERENCES LIST FOR CHAPTER 4

[590] C. C. Doumanidis and D. E. Hardt. Multivariable adaptive control of thermal properties during welding. Proceedings of the Winter Annual Meeting of the ASME on Control Methods for the Manufacturing Processes. Chicago, IL. Editor D. E. Hardt. Nov.-Dec., 1-12, 1988.

[591] C. C. Doumanidis and D. E. Hardt. A model for in-process control of thermal properties during welding. Transactions of the ASME, Journal of Dynamic Systems,Measurement, and Control. Volume 111, Pages 40-50. March, 1989.

[592] C. C. Doumanidis. Hybrid modeling for control of weld dimensions. Japan/USA Symposium on Flexible Automation. Pages 317-323, 1992.

[593] C. C. Doumanidis. Multiplexed and distributed control of automated welding. IEEE Control Systems. Volume 14(4), Pages 13-24. August, 1994.

[594] C. Doumanidis. Modeling and control of timeshared and scanned torch welding. Transactions of ASME, Journal of Dynamic System, Measurement and control. Volume 116, Pages 387-394. September, 1994.

[595] N. Fourligkas and C. Doumanidis. Distributed parameter control of automated welding processes. Proceedings of 2nd IEEE Mediterranean Symposium On New Direction in Control & Automation. Pages 113-119. Maleme-Chania, Crete, Greece. June 19-22, 1994.

[596] J. F. Collard. Adaptive pulsed GMAW control: The digipulse system. Welding Journal. Volume67(11), Pages35-38. November, 1988.

[597] G. E. Cook, K. Andersen and R. J. Barrett. Welding and bonding. The Electrical Engineering Hand Book. CRC Press. Boca Raton, FL. R. C. Dorf. Pages 2223-2237, 1993.

[598] W. A. Wolovich. Linear Multivariable Systems. Springer-Verlag. New York, NY., 1974.

[599] W. M. Wonham. Linear Multivariable Control: A Geometric Approach, Third Edition. Springer-Verlag. New York, NY., 1979.

[600] M. O. Tade and D. W. Bacon. Adaptive decoupling of a class of multivariable dynamic systems using output feedback. IEE Proceedings. Vol 133, No 6, 1986.

[601] Y. M. Kwak and C. Doumanidis. Solid freeform fabrication by GMA welding: Geometry modeling, adaption and control. American Society of Mechanical Engineers, Manufacturing Engineering Division. Volume10, Pages49-56, 1999.

[602] T. O. Santos, R. B. Caetano, J. M. Lemos and F. J. Coito. Multipredictive adaptive control of arc welding trailing centerline temperature. IEEE Transactions on control Systems Technology. Volume8(1), Pages159-169. January, 2000.

[603] K.L. Moore, D.S. Naidu, M. Abdelrahman and A. Yesildirek. Advanced Welding Control Project: Annual Report FY96. Measurement and Control Engineering Research Center. Idaho State University, Pocatello, ID. June 28, 1996.

[604] S. Ozcelik. Design of Robust Feed forward Compensators for Direct Model Reference Adaptive Controllers. Rensselaer Polytechnic Institute. Troy, NY., 1996.

[605] S. Ozcelik, K. L. Moore and D. S. Naidu. Adaptive control of a gas metal arc welding (GMAW) process. Measurement and Control Engineering Research Center, Idaho State University. Pocatello, ID. April, 1997.

[606] M. Janots. Adaptive system of resistance welding control. Fourth International Conference on Advances in Welding Processes. Harrogate, England. May. Pages239-247, 1978.

[607] G. E. Cook. Feedback and adaptive control of process variable in arc welding. Proceedings of an International Conference on Developments in Mechanized, Automated and Robotic Welding. Welding Institute. Cambridge, MA. Pages P321-P329. November, 1980.

[608] G. E. Cook. Feedback and adaptive control in automated arc welding systems. Metal Construction. Pages 551-556. Volume 13(9). September, 1981.

[609] G. E. Cook, K. Andersen and R. J. Barnett. Feedback and adaptive control in welding. Recent Trends in Welding Science and Technology. Metals Park, OH. S. A. David and J. M. Vick. Pages 891-903, 1990.

[610] G. E. Cook. Microcomputer control of an adaptive positioning system for robotic arc welding. Proceedings of IEEE IECI Conference on Applications of Mini and Microcomputers. San Francisco, CA. Pages 324-329. November, 1981.

[611] W. J. Kerth, Sr. and W. J. Kerth, Jr.. Digitally-controlled adaptive welding system. Proceedings of Tenth NSF Conference on Production Research Technology. Pages 139-140. Feb./March., 1983.

[612] G. Nachev, B. Petkov, L. Blagoev and I. Tsankarski. Adaptive gas metal arc GMA welder. Proceedings of th SPIE 3rd International Conference on Robot Vision and Sensory Controls, RoViSeC3. Cambridge, MA. Volume449, Pages286-290. November, 1983.

[613] L. M. Sweet, A. W. Case Jr, N. R. Corby and N. R. Kuchar. Closed-loop joint tracking, puddle centering and weld process control using an integrated weld torch vision system. Proceedings of the ASME Winter Annual Meeting on Control of Manufacturing Processes and Robotic Systems. Boston, MA. D. E. Hardt and W. J. Book. Pages 97–105. November, 1983.

[614] K. A. Kuk. Determining acceptable joint mislocation in systems without adaptive control. Welding Journal. Pages65-66. November, 1985.

[615] M. A. Burke, H. B. James and R. M. Wells. The robotic adaptive welding system-RAWS. Proceedings of First International Conference on Advanced Welding systems. Houldcroft, P.T. Welding Institute, Cambridge, UK. London, UK. Pages 31-40. November, 1987.

[616] K. Haefner, B. Carey, B. Bernstein, K. Overton and M. D'Andrea. Real time adaptive spot welding control. Proceedings of the Winter Annual Meeting of the ASME on Control Methods for the Manufacturing Processes. Chicago, IL. D. E. Hardt. Nov.-Dec.. Pages 51-62, 1988

[617] A. Suzuki and D. E. Hardt. Application of adaptive control theory to on-line GTA weld geometry regulation. Proceedings of the Winter Annual Meeting of the ASME on Control Methods for the Manufacturing Processes. Chicago, IL. D. E. Hardt. Nov.-Dec.. Pages 13-25, 1988

[618] A. Suzuki, D. E. Hardt and L. Valavani. Application of adaptive control theory to on-line GTA weld geometry regulation. Transactions of the ASME Journal of Dynamic Systems, Measurement and Control. Volume 113, Pages 93–103. March, 1991.

[619] K. S. Narendra and Y. Lin. Stable discrete adaptive control. IEEE Transactions on Automatic Control. Volume AC-25, June, 1980.

[620] K. J. Astrom and B. Wittenmark. Self-tuning controllers based on pole-zero placement. IEEE Proceedings. Volume 127. May, 1980.

[621] J. L. Pan, R.H. Zhang, Z. M. Ou, Z. Q. Wu and Q. Chen. Adaptive control GMA welding - a new technique for quality control. Welding Journal. Volume 68(3), Pages 73-76. March, 1989.

[622] C. Conrardy. Control of GMAW with coaxial vision. Ohio State University, 1991.

[623] P. Drews and A. H. Kuhne. An automatic welding system. Proceedings of EWI First International Conference on Advanced Welding Systems. P. T. Houldcroft. Edison Welding Institute. London, UK. Pages 191-198. November, 1985.

[624] H. Nomura. Basics for welding process control. Sensors and Control Systems in Arc Welding. Chapman & Hall. London, UK. English Translation of the Original 1991 Japanese Edition. Chapter 4, Pages 53-68. 1994.

[625] K. Fujita and T. Ishide. H. Nomura. Adaptive control of welding conditions using visual sensing. Sensors and Control Systems in Arc Welding. Chapman & Hall. London, UK. English Translation of the Original 1991 Japanese Edition. Chapter 18, Pages 160-167, 1994.

[626] L. J. Brown and S. P. Meyn. Adaptive dead-time compensation. Proceedings of the IEEE Conference on Decision and Control. Volume4. New Orleans, LA. Pages3435-3437, 1995.

[627] J. Brown, S. P. Meyn and R. A. Weber. Adaptive dead-time compensation with applications to a robotic arc welding system. University of Illinois. Urbana, IL., 1995.

[628] Y. M. Zhang, B. L. Walcott and L. Wu. Adaptive predictive decoupling control of full penetration process in GTAW. Proceedings of the 1st IEEE Conference on Control Application. Dayton, OH. Pages 938-943, 1992.

[629] Y. M. Zhang, R. Kovacevic and L. Wu. Closed-loop control of weld penetration using front-face vision sensing. Proceedings of Institution of Mechanical Engineers Part I: Journal of Systems and Control Engineering. Volume 207, Pages 27-34, 1993.

[630] R. Kovacevic and Y. M. Zhang. Model-based adaptive vision control of weld pool area. Proceedings of the American Control Conference. Baltimore, MD. June. Pages 313-317, 1994.

[631] R. Kovacevic and Y. M. Zhang. Adaptive control of GTA welding of sheet metals. Proceedings of the Fifth International Welding Computerization Conference. Golden, CO. Pages 38-48. August, 1994.

[632] E. Mosca. Optimal Predictive, and Adaptive Control. Prentice Hall. Englewood Cliffs, NJ., 1995.

[633] Y. M. Zhang, R. Kovacevic and L. Li. Adaptive control of full penetration gas tungsten arc welding. IEEE Transaction on Control Systems Technology. Volume 4(4), Pages 394-403. July, 1996.

[634] R. Kovacevic and Y. M. Zhang. Sensing free surface of arc weld pool using specular reflection: principle and analysis. Proceedings of Institution of Mechanical Engineers, Part B of Engineering Manufacturing. Volume 210(6), Pages 553-564, 1996.

[635] Y. M. Zhang, R. Kovacevic and L. Wu. Controlling welding penetration in GTAW using vision sensing and adaptive control technique. ASME Transactions. Volume XX. Pages 317-324, 1992.

[636] Y. M. Zhang, R. Kovacevic and L. Wu. Robust adaptive control of full penetration using weld depression feedback. Proceedings of the Fifth International Welding Computerization Conference. Golden, Colorado. Pages 99-110. August 10-12, 1994.

[637] Y. M. Zhang, Y. X. Yao, L. Li and R. Kovacevic. Based robust multivariable linear control of weld pool shape. Proceedings of the ASME Dynamic Systems Control Division. ASME International Mechanic Engineering Congress. Atlanta, GA. Volume 58, Pages 259-264. November 17-22, 1996.

[638] Y. Sugitani and W. Mao. Automatic simultaneous control of bead height and back bead shape using an arc sensor in one-sided welding with a backing plate. Welding Research Abroad. Vol 42, No 12, Pages 9-17. December, 1996.

[639] S. G. Tzafestas, G. G. Rigatos and E. J. Kyriannakis. Geometry and thermal regulation of GMA welding via conventional and neural adaptive control. Journal of Intelligent and Robotic Systems. Volume19(2), Pages153-186. June, 1997.

[640] Poolsak Koseeyaporn, E. Geroge Cook and M. Alvin Strauss. Adaptive voltage control in fusion arc welding. IEEE Transactions on Industry Applications. Volume36(5), Pages1300-1307. September, 2000.

[641] L. A. Zadeh. Fuzzy sets. Information and Control. Volume 8, Pages 338-353, 1965.

[642] L.-X. Wang. A Course in Fuzzy Systems and Control. Prentice Hall PTR. Upper Saddle River, NY., 1997.

[643] G. Langari and M. Tomizuka. Fuzzy linguistic control of arc welding. Proceedings of ASME Winter Annual Meeting on Sensors and Controls for Manufacturing. Chicago, IL. Pages 157-162. November, 1988.

[644] K. Ohshima, S. Yamane, H. Iida, S. Xiang, Y. Mori and T. Kobota. H. Nomura. Fuzzy control of CO_2 short-arc welding. Sensors and Control Systems in Arc Welding. Chapman & Hall. London, UK. English Translation of the Original 1991 Japanese Edition. Chapter 15, Pages 137-146, 1994.

[645] H. Fujimura, E. Ide and H. Inoue. H. Nomura. Robot welding with arc sensing. Sensors and Control Systems in Arc Welding. Chapman & Hall. London, UK. English Translation of the Original 1991 Japanese Edition. Chapter 26, Pages 228-237, 1994.

[646] Z. Bingul and G. Cook. Dynamic modeling of GMAW process. Proceedings - IEEE International Conference on Robotics and Automation. Piscataway, New Jersey. Pages3059-3064, 1999.

[647] H. B. Smartt and J. A. Johnson. Intelligent control of arc welding. EG&G, Idaho National Engineering Laboratory. Idaho Falls, ID. September, 1989. Preprint.

[648] H. S. Cho. Application of AI to welding process automation. Proceedings of ASME Japan/USA Symposium Flex. Auto. Pages 303-308. July, 1992.

[649] H. B. Smartt, J. A. Johnson, C. J. Einerson, A. D. Watkins and N. M. Carlson. Model-based approach to intelligent control of gas metal arc welding. Proceedings of the 5th International Conference on Modeling of Casting, Welding and Advanced Solidification Processes. Davos, Switzerland. September. M. Rappaz, M. R. Ozgu and K. W. Mahin. Pages 305–313, 1990.

[650] H. B. Smartt and J. A. Johnson. A novel implementation of fuzzy logic – theory type Unpublished Technical Report. EG&G, Idaho, Idaho National Engineering Laboratory. Idaho Falls, ID. April, 1991.

[651] H. B. Smartt. Intelligent sensing and control of arc welding. Proceedings of the ASM 3rd International Conference on Trends in Welding Research. Gatlinburg, TN. S. A. David and J. M. Vitek. Pages843-851. Keynote Address. June, 1992.

[652] S. B. Chen, L.Wu, Q. L. Wang and Y. C. Liu. Self - learning fuzzy neural networks and computer vision for control of pulsed GTAW. Welding Journal. Pages 201s-209s. May, 1997.

[653] R. Kovacevic and Yu M. Zhang. Neurofuzzy model-based weld fusion state estimation. IEEE Control Systems. Pages 30-42. April, 1997.

[654] J.-S. R. Jang, C.-T Sun, and E. Mizutani. Neuro-Fuzzy and Soft Computing. Prentice Hall, Upper Saddle River, NJ, 1997.

[655] Y M. Zhang, R. Kovacevic and L. Wu. Sensitivity of front-face weld geometry in representing the full penetration. Proceedings of Institution of Mechanical Engineers, Part B: Journal of Engineering Manufacture. Volume 206(3), Pages 191-197, 1992.

[656] Y. M. Zhang, L. Wu, B. L. Walcott and D. H. Chen. Determining joint penetration in GTAW with vision sensing of weld-face geometry. Welding Journal. Volume 72. October. Pages 463s-469s, 1993.

[657] Y. M. Zhang and R. Kovacevic. Modeling and real-time identification of weld pool characteristics for intelligent control. The First Congress on Intelligent Manufacturing Process & Systems, CIRP. Mayaguez, Puerto Roco. February 13-17, 1995.

[658] Y. M. Zhang, R. Kovacevic and L. Wu. Dynamic analysis and identification of gas tungsten arc welding process for full penetration control. Transactions of ASME, Journal of Engineering for Industry. Volume 118. Pages 123-136. February, 1996.

[659] R. Kovacevic, Y. M. Zhang, and S. Ruan. Sensing and control of weld pool geometry for automated GTA welding. ASME Transactions, Journal of Engineering for Industry. Vol 117, No 2, Pages 210-222, 1995.

[660] Y. M. Zhang, L. Li and R. Kovacevic. Dynamic estimation of full penetration using geometry of adjacent weld pool. ASME journal of Manufacturing Science and Engineering. Volume 119(2). May, 1997.

[661] Y. M. Zhang and R. Kovacevic. Real-time sensing of sag geometry during GTA welding. ASME Journal of Manufacturing Science and Engineering. Volume 119(2), Pages 1-10. May, 1997.

[662] K. Ohshima, S. Yamane, Y. Mori and P. Ma. H. Nomura. Application of a fuzzy neural network to welding line tracking. Sensors and Control Systems in Arc Welding. Chapman & Hall. London, UK. English Translation of the Original 1991 Japanese Edition. Chapter 16, Pages 147-1153, 1994.

[663] Li Di, Song Yonglun and Ye Feng. On Line monitoring of weld defects for short-circuit gas metal arc welding based on the Self-Organize feature Map neural networks. Proceedings of the International Joint Conference Neural Networks. Piscataway, New Jersey. Pages239-244. July, 2000.

[664] J. E. Agapakis, K. Masubuchi and N. Wittels. General visual sensing techniques for automated welding fabrication. Proceedings of the 4th International Conference on Robot Vision and Sensory Controls RoViSeC4. London, UK. Pages 103-114. October, 1984.

[665] J. E. Agapakis and J. M. Katz. An approach for knowledge-based sensor-controlled robotic welding. Proceedings of the 2nd International Conference on Trends in Welding Research. S. A. David and J. M. Vitek. Gatlinburg, TN. Pages935-939. May, 1989.

[666] A. H. Kuhne, H.B. Cary and F. B. Prinz. An expert system for robotic arc welding. Welding Journal. Pages21. November, 1987.

[667] H. Watanabe, Y. Kondo and K. Inoue. H. Nomura. Automatic control technique for narrow-gap GMA welding. Sensors and Control Systems in Arc Welding. Chapman & Hall. London, UK. English Translation of the Original 1991 Japanese Edition. Chapter 21. Pages 182-190, 1994.

[668] R. E. Reeves, T. D. Manley, A. Potter, and D. R. Ford. Expert system technology- an avenue to an intelligent weld process control system. Welding Journal. Pages 33-41. June, 1988.

[669] S. Misra, S. Subramanyam, S. Pandey and P. B. Sharma. Expert system for GMA welding of aluminum. Proceedings of ASM 3rd International Conference on Trends in Welding Research. S. A. David and J. M. Vitek. Gatlinburg, TN. Pages949-956. June, 1992.

[670] U. Dilthey and J. Heidrich. Using AI-methods for parameter scheduling, quality control and weld geometry determination in GMA-welding. ISIJ International. Volume 39(10), Pages1067-1074, 1999.

[671] G. E. Cook. Through-the-arc sensing for arc welding. Proceedings of the 10th NSF Conference on Production Research and Technology. Pages 141-151. February28-March2, 1983.

[672] K. Andersen, R. J. Barnett and G. E. Cook. Intelligent Gas Tungsten Arc Welding: Final Report. Vanderbilt University. Nashville, TN., 1988.

[673] G. E. Cook, K. Andersen, R. J. Barnett, and J. F. Springfield. Intelligent gas tungsten arc welding. Automated Welding Systems in Manufacturing. J. Weston. Cambridge, UK., 1991.

[674] J. Doherty, S. J. Holder and R. Baker. Computerized guidance and process control. Proceedings of Third International Conference on Robot Vision and Sensory Controls RoViseC3. Cambridge, MA. Volume 449. D. P. Cassasent and E. L. Hall. Pages 482-487. November, 1983.

[675] D. R. White and J. E. Jones. A hybrid hierarchical controller for intelligent control of welding processes. Proceedings of the ASM 2nd International Conference on Trends in Welding Research. Gatlinburg, TN. May. S. A. David and J. M. Vitek. Pages 909-915, 1989.

[676] D. R. White, J. A. Carmein, J. E. Jones and K. Liu. Integration of process and control models for intelligent control of welding.

Proceedings of the ASM 3rd International Conference on Trends in Welding Research. Gatlinburg, TN. June. S. A. David and J. M. Vitek. Pages 883-887, 1992.

[677] T. Kangsanant and Z.H. Wang. CAD-based expert system for robotic welding. Proceedings of the 12th Triennial World Congress of the International federation of Automatic Control. Pages 303-306. Sydney, Australia. July, 1993.

[678] Anonymous. Computer-based smart processes designed to increase U.S. manufacturer's quality. Welding Journal. Volume75. October. Pages18-19, 1996.

[679] T. A. Siewert, R. B. Madigan and T. P. Quinn. Sensors Control gas metal arc welding. Advanced Materials & Processes. Volume151(4), Pages23-25. April, 1997.

[680] U. Dilthey, J. Heidrich and T. Reichel. On-Line quality control in gas metal arc welding using artificial neural networks. Schweissen und Schneiden/ Welding & Cutting. Volume49(2), Pages E22-E24. February, 1997.

[681] N. Yetukuri and G. W. Fischer. Planning the GMAW process by constraint propagation. Journal of Intelligent Manufacturing. Volume8(6), Pages477-488. December, 1997.

[682] P. Verdelho, M. Pio Silva, E. Margato and J. Esteves. Electronic welder control circuit. IECON Proceedings (Industrial Electronics Conference), Proceedings of the 1998 24th Annual Conference of the IEEE Industrial Electronics Society. Los Alamos, California. August. Pages612-617, 1998.

[683] L. Brown, S. Meyn and R. Weber. Adaptive dead-time compensation with application to a robotic welding system. IEEE Transactions on Control Systems Technology. Volume6(3), Pages335-349. May, 1998.

[684] K.-J. Matthes, M. Kusch, H. Roth. S. Mueller and F. Wuest. Control response of pulsed power sources for gas-shielded metal-arc welding when the contact-tube distance is changed. Schweissen

und Schneiden / Welding and Cutting. Volume 51(9), Pages201-204, 1999.

[685] S. J. Park and T. J. Lim. Fault-tolerant robust supervisor for discrete event systems with model uncertainty and its application to a workcell. IEEE Transactions on Robotics and Automation. Volume 15(2), Pages 386-391, 1999.

[686] G. Cook. Robotic Arc Welding. Technical Paper - Society of Manufacturing Engineers. (AD99-109), Pages 1-20. 1999.

[687] C. S. Wu, T. Polite and D. Rehfeldt. A Fuzzy Logic System for Process Monitoring and Quality Evaluation in GMAW. Welding Journal. Pages 33s-38s. February, 2001.

[688] I. W. Barr. Elementary Statistical Quality Control. Dekker Publications. New York, NY., 1979.

[689] C. Peterson. Statistical Process Control. Viking Press Inc. Waterloo, Iowa, 1987.

[690] J. E. Maxwell. Statistical Process Control for Arc Welding Processes. Vanderbilt University. Nashville, TN,1990.

[691] G. E. Cook, R. J. Barnett, A. M. Strauss and F. M. Thompson Jr. Statistical weld process monitoring with expert interpretation. Proceedings of the ASM 4th International Conference on Trends in Welding Research. Gatlinburg, TN. H. B. Smartt, J. A. Johnson and S. A. David. Pages 739-744. June 5-8, 1995.

[692] G. E. cook, J. E. Maxwell, R. J. Barnett and A. M. Strauss. Statistical process control application to weld process. IEEE Transactions on Industry Applications. Volume33(2), Pages 454-463. March/April, 1997.

[693] H. M.Wadsworth. Handbook of Statistical Methods for Engineers and Scientists. McGraw Hill. New York, NY., 1990.

[694] P Gary Maul, Richard Richardson and Brett Jones. Statistical process control applied to gas metal arc welding. Computers And Industrial Engineering. Volume 31(1-2), Pages 253-256. October, 1996.

[695] S.V. Dubovetskii. The application of statistical models of weld formation to the design and control of automatic arc welding conditions. Proceedings of the EWI 1st International Conference on Advanced Welding System. London, UK. November. P. T. Houldcroft. Pages 210-236, 1985.

[696] P. L. Cook. Quality control systems for pipeline welding-a model and quantitative analysis. Welding Journal. Pages 39-47. March, 1985.

[697] S. I. Rokhlin, K. Cho and A. C. Guu. Closed-loop process control of weld pool penetration using real-time radiography. Materials Evaluation. Volume47, Pages 363-369. March, 1989

[698] J. C. Papritan and S. C. Helzer. Statistical process control for welding. Welding Journal. Pages 44-48. March, 1991.

[699] R. Reilly. Real-time weld quality monitor control GMA welding. Welding Journal. Pages 36-41. March, 1991.

[700] G. P. Maual, R. Richardson and B. Jones. Statistical process control applied to gas metal arc welding. Computers and Industrial Engineering. Volume31. 1-2. Pages 253-256. October, 1996.

[701] J. I. Lee and S. Rhee. Prediction of process parameters for gas metal arc welding by multiple regression analysis. Proceedings of the Institution of Mechanical Engineers, Part B: Journal of Engineering Manufacture. London, England. Pages 443-449, 2000.

[702] M. Lanzetta, M. Santochi and G. Tantussi. On-line Control of robotized gas metal arc welding. CIRP Annals - Manufacturing Technology. Volume 50(1), Pages 13-16, 2001.

[703] K. L. Moore. Iterative Learning Control - An Expository Review. Idaho State University. Pocatello, ID. T.R. 96/97 003. June, 1997.

[704] K. L. Moore and A. Mathews. Iterative learning control of systems with non-uniform trial length with applications to gas metal arc welding. Proc. of the 2nd Asian Control Conference. Seoul, Korea. July, 1997.

[705] J.-J. Slotine and W. Li. Applied Nonlinear Control. Prentice Hall. Englewood Cliffs, NJ., 1991.

[706] A. Isidori. Nonlinear Control Systems, Third Edition. Springer-Verlag, Berlin, 1995.

[707] K.L. Moore, M.A. Abdelrahman and D.S. Naidu. Gas metal arc welding control: part II- control strategy. Proceedings of the Second World Congress of Nonlinear Analysis. Athens, Greece. July, 1996.

[708] K. L. Moore, M. A. Abdelrahman and D. S. Naidu. Gas metal arc welding control: Part II - control strategy. Nonlinear World. August, 1997.

[709] K. L. Moore, M. A. Abdelrahaman and D. S. Naidu. Gas metal Arc welding control - II Control Strategy. Nonlinear Analysis, Theory, Methods & Applications. Volume 35(1), Pages 85-93. January, 1999.

[710] T. E. Martin. Process Control: Designing Processes and Control Systems for Dynamic Performance. McGraw Hill. New York, NY., 1995.

[711] J. Tyler. Model-Based Control of Gas Metal Arc Welding Process. Idaho State University. Pocatello, ID, 1997.

[712] J. J. Hunter, G. W. Bryce and J. Doherty. On-line control of the arc welding process. Proceedings of the International Conference on Developments in Mechanized Automated and Robotic Welding Process. London, UK. The Welding Institute. Pages P37-1-P37-11. November, 1980.

[713] H. Robinson and N. R. Nutter. The development of automatic welding techniques for nuclear applications. Proceedings of an International Conference on Developments in Mechanized Automated and Robotic Welding. London, UK. November. Pages P31-P315, 1980.

[714] G. E. Cook, P.C. Levick, D. Welch and A.M. Wells, Jr. Distributed microcomputer control of an automated arc welding system. IEEE IAS Conference. Pages 1296-1302. October, 1982.

REFERENCES LIST FOR CHAPTER 4

[715] W. A. Heller. Fundamentals and capabilities of automated welding systems. Proceedings of the Conference on Automation and Robotics for Welding. Indianapolis, IN. Pages 17-29. February, 1983.

[716] M. Kawahara. Tracking control system for complex shape of welding groove using image sensor. Proceedings of IFAC/IFIP Symp.- Real Time Digital Control Applications. Guadalajara, Mexico. Pages 257-263. Jan, 1983.

[717] N. R. Corby Jr. Machine vision algorithms for vision guided robotic welding. Proceedings of 4th International Conference on Robot Vision and Sensory Controls, RoViseC5. London, UK. Pages 137-147. October, 1985.

[718] S. Habib. Process control package designs, monitors, evaluates welding operation. Welding Journal. Pages 69-70. March, 1987.

[719] Z. Smati, P. J. Alberry and D. Yapp. Strategies for automatic multipass welding. Proceedings of First International Conference on Advanced Welding System. P. T. Houldcroft. Volume 519, Pages 219–237. London, UK. November, 1987.

[720] W. R. Schick. Verticle strip cladding: process control. welding Journal. Pages 17-22. March, 1988.

[721] K. Andersen, G. E. Cook, R. J. Barnett and E. H. Eassa. A class-H amplifier power source used as a high-performance welding research tool. Proceedings of the 2nd International Conference on Trends in Welding Research. S. A. David and J. M. Vitek. Gatlinburg, TN. Pages 973-978. May, 1989.

[722] J. B. Bjorgvinsson. Adaptive Voltage Control in Gas Tungsten Arc Welding. Vanderbilt University. Nashville, TN, 1992.

[723] K. N. Groom, S. Nagarajan and B. A. Chin. Automatic single V groove welding utilizing infrared images for error detection and correction. Welding Journal. Pages 441s-445s. December, 1990.

[724] G. E. Cook. Computer-based control system for gas tungsten arc welding. Japan/USA Symposium on Flexible Automation. Pages 197-301, 1992.

[725] W. S. Jaffery. Automated robotic variable polarity plasma arc welding (VPPAW) for the space station Freedom project (SSFP). Proceedings of the 3rd International Conference on Trends in Welding Research. S. A. David and J. M. Vitek. Gatlinburg, TN. Pages 943-947. June, 1992.

[726] D. E. Hardt. Modeling and control of manufacturing processes: getting more involved. Transactions of the ASME, Journal of Dynamic Systems, Measurement and control. Volume115. Pages 291-300. June, 1993.

[727] H. Nomura. Sensors and Control Systems in Arc Welding. Chapman & Hall. London, UK. English Translation of the Original 1991 Japanese Edition, 1994.

[728] H. Nomura. Control systems. Sensors and Control Systems in Arc Welding. Chapman & Hall. London, UK. English Translation of the Original 1991 Japanese Edition. Chapter 3. Pages 44-52, 1994.

[729] L. Dorn, K. Moneni, H. Mecke and T. Rummel. Process control in the arcing and short-circuit phases during metal-arc active gas build-up welding under carbon dioxide. Welding and Cutting. Volume 41(6-7), Pages 2-5. June-July, 1994.

[730] H. Nomura. Future trends. Sensors and Control Systems in Arc Welding. Chapman & Hall. London, UK. English Translation of the Original 1991 Japanese Edition. Chapter 7. Pages 86-87, 1994.

[731] H. Holm. Manufacturing state modeling and its applications. Proceedings of the ASM 4th International Conference on Trends in Welding Research. Gatlinburg, TN. H. B. Smartt and J. A. Johnson and S. A. David. Pages 665-675. June 5-8, 1995.

[732] A. E. Bentley and S. J. Marburger. Arc welding penetration control using quantitative feedback theory. Welding Journal. Volume71(11), Pages 397-404. June 5-8, 1992.

[733] J. J. D'Azzo and C. H. Houpis. Linear Control System Analysis and Design: Conventional and Modern. Edition Fourth. McGraw-Hill. New York, NY, 1995.

REFERENCES LIST FOR CHAPTER 4

[734] H. B. Smartt, J. A. Johnson and S. A. David. Proceedings of the ASM 4th Int. Conf. on Trends in Welding Research. ASM International. Gatlinburg, TN. June, 1995.

[735] P. Zhu, M. Rados and S. W. Simpson. Theoretical predictions of the start-up phase in GMA welding. Welding Journal. Volume 76(7), Pages 269s-274s. July, 1997.

[736] R. Hajossy, P. Pastava and I. Morva. Ignition of a welding arc during a short-circuit of melted electrodes. Journal of Physics D: Applied Physics. Volume 32(9), Pages 1058-1065, 1999.

[737] D. A. Hartman, D. R. Delapp, G. E. Cook and R. J. Barnett. Intelligent control in arc welding. Intelligent Engineering Systems Through Artificial Neural Networks. Volume 9, Pages 715-725, 1999.

[738] G. S. Tzafestas and J. E. Kyriannakis. Regulation of GMA welding thermal characteristics via a hierarchical MIMO predictive control scheme assuring stability. IEEE Transactions on Industrial Electronics. Volume 47(3), Pages 668-678, 2000.

[739] Y. M. Zhang and E. Liguo. Numerical analysis of dynamic growth of droplet in GMAW. Journal of Mechanical Engineering Science. Pages 1247-1258, 2000.

[740] Y. M. Zhang, Liguo E. and B. L. Walcott. Robust control of pulsed gas metal arc welding. ASME journal of Dynamic Systems, Measurement and Control. May, 2001.

[741] Y. M. Zhang and P. J. Li. Modified active control of metal transfer and pulsed GMAW of titanium. Welding Journal. Pages 54s-61s, 2001.

[742] T. Eagar, D. Hardt and J. Lang. Decoupling of thermally based processes for multivariable control. Part of Joint INEL, MIT, ISU Proposal submitted to DOE Office of Basic Energy Sciences, 1993.

[743] J. L. Schiano. Feedback control of two physical processes: design and experiments. University of Illinois, Urbana, IL. No DC-148, UILU-ENG-93-2217, 1993.

[744] M. E. Shepard. Modeling of Self-Regulation in Gas Metal Arc welding. Vanderbilt University. Nashville, TN, 1991.

[745] M. E. Shepard and G. E. Cook. A frequency-domain model of self-regulation in gas metal arc welding. Proceedings of the ASM 3rd International Conference on Trends in Welding Research. Gatlinburg, TN. S. A. David and J. M. Vitek. Pages 899-903. June, 1992.

[746] P. J. Hewitt and A. A. Hirst. A systems approach to the control of welding fumes at source. Annals of Occupational Hygiene. Volume 37, Pages 297-306. June, 1993.

Chapter 5

Control of GMAW: A Case Study

In this chapter, we present a case study on the control strategy for the Gas Metal Arc Welding (GMAW) process that was performed by the authors and their students at Idaho State University (ISU), Pocatello, Idaho. The work was performed in collaboration with the researchers at the Idaho National Engineering and Environmental Laboratory (IN-EEL).

5.1 Introduction

The research presented here is based on the following chain of ideas. Generally, a good weld is identified by its microstructure and other factors (e.g., the amount of spatter, the amount of overfill or underfill, etc.). Although these are not easily measured or quantified, they can be related to characteristics such as the cooling rate of the weld pool, the metal transfer mode, the bead/groove geometry, workpiece defects, etc. Likewise, many of these characteristics can be related to the mass and heat transferred from the GMAW process to the weld pool. These are affected in turn by the properties of the stream of droplets that are melting off the electrode. Typically, we want the stream of droplets to be uniform in size and come off the electrode at uniform intervals. If a suitable degree of control authority over the droplet properties (detachment interval and droplet size) can be achieved, it would be possible to control the heat and mass transfer and, consequently, the "goodness" of the weld. Following these ideas, our approach to controlling the quality of the weld in the GMAW process is to adjust the power supply parameters (perhaps based on measurements of process conditions) so as to control the heat and mass input to the weld pool and, more specifically, the droplet detachment properties.

Figure 5.1 gives a pictorial description of the interdisciplinary nature of the GMAW control problem and also illustrates the general approach we have taken in our long-term research on GMAW automation [747, 748]. The figure also shows an input/output model of the GMAW process as well as the primary signals in the system. We consider the process to have four inputs (open-circuit voltage, wire-feed speed, contact tube-to-workpiece distance, and torch travel speed, or V_{oc}, S, CT, and R, respectively) and two measured outputs (current and arc voltage (shown equivalently as arc length in the figure), or I and V_{arc}, respectively). These primary signals act to produce the heat and mass inputs to the weldment (H and G, respectively), which then interact with the weldment to produce the thermal and fill properties of the weld. We additionally assume that sensors and analysis techniques exist to measure various properties of the weld pool geometry (e.g., seam tracking coordinates or required fill) and thermal properties (e.g., centerline cooling rate or penetration). Using the information from the available sensors, the task of the controller is to select the best values of the process inputs so as to produce "good" welds.

The results we present are part of the larger project aimed at developing methods for controlling GMAW processes, as depicted in Figure ??. The long-range emphasis is on the control of the mass and heat delivered by the process to the weld pool [749, 750]. However, as a part of the larger project, the variables closest to the process were chosen as a starting point, with a focus on regulation of the measured current and arc voltage to desired set points. Experimental results showing the ability to achieve such regulation using classical controller design techniques are presented. Also included are the results of the application of more modern adaptive control techniques to the GMAW control problem, specifically, direct model reference adaptive control (DMRAC).

Accurate modeling is an important part of any control system de-

5.1. INTRODUCTION

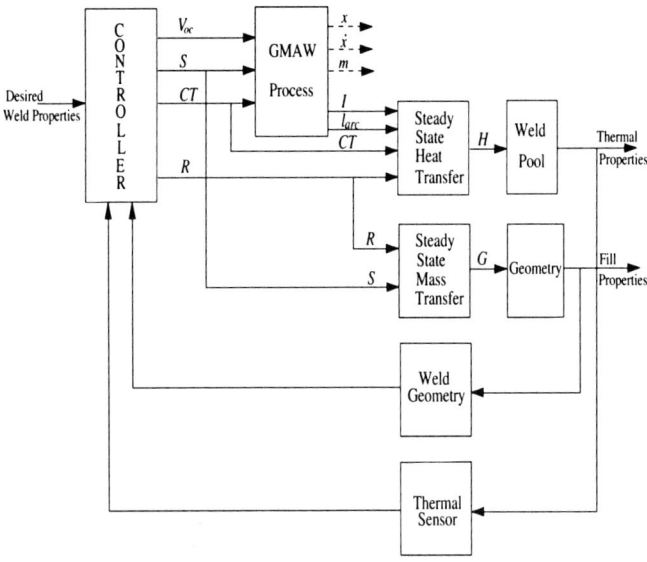

Figure 5.1: The "big-picture" of GMAW process control.

sign, regardless of the algorithms used in the controller. To understand the process dynamics, experiments were conducted to calibrate a fifth-order, nonlinear dynamic model of the GMAW process for a specific experimental apparatus. The fifth-order model is based on past research and is derived from a combination of first principles and empirical results available in the literature [751, 748, 752]. A key feature of the model is that it is parameterized by various physical constants and in particular by five empirical constants. Experimental data of steady-state current and arc voltage was obtained from an automated GMAW system over a range of operational conditions defined by variations in open-circuit voltage and wire-feed speed. The data is used to characterize the optimal values of the five empirical constants used in the model where best is defined relative to the error between the actual GMAW outputs and those predicted by the model. Consideration is given to the optimization of parameters over the complete data set. The analysis is not unlike that in [751], which found the best values of empirical parameters from a large data set collected over a wide range of operating points. However, here the optimal value of each parameter is determined as a function of operating conditions. From contour plots of the model error, it is seen that the empirical "constants" of the fifth-order model are not constants but instead are a function of the operating conditions of the system. This is a key observation that motivates the use of the DMRAC later in the chapter.

In previous works, it was argued that it is reasonable to approximate the GMAW process with a two-by-two linear, multivariable system [752, 748]. From simple step response tests, it is possible to derive an empirical model in transfer function form for nominal operating conditions. To this end, an empirically-determined dynamic model of the GMAW process for use in the controller design is described and model analysis, controller design, and experimental results for two controller strategies were presented. First, a single-input, single-output (SISO) proportional integral (PI) controller for current is given. Next, the relative gain array (RGA) technique of process control is applied to the empirical model to design a multi-loop PI controller for the process. The resulting controller pairs wire-feed speed with current and open-circuit voltage with arc voltage to regulate current and arc voltage to desired set points. Using the error between the measured values of current and arc voltage and the desired values of these variables, the controller simultaneously adjusts the wire-feed speed and the open-circuit voltage of the power supply, respectively. The basic benefit that is derived at this stage is the ability to reduce variability in the measured signals combined with the ability to force the measured outputs to their desired values. One of the distinguishing features of much of the work has been the model-based approach to the design of the controllers for the process [752]. Also, the use of the RGA method to select controller loop pairings is unique and offers interesting insight into the best ways to control the process [753].

5.2 Empirical Modeling of a GMAW Process

The theoretical modeling of the GMAW process used to derive the INEEL/ISU model was given in Section 2.9. In this section, we describe the empirical determination of two related models. First, we present the calibration of the fifth-order nonlinear model, including a description of the data collected and the parameter optimization re-

5.2. EMPIRICAL MODELING OF A GMAW PROCESS

sults. We then present an empirically derived transfer function model of the process about a specific operating point.

5.2.1 Calibration of a GMAW Process

Experimental Data

Using an experimental facility constructed to investigate welding control and automation, a series of 55 experiments were conducted to collect calibration data. In each experiment, a prescribed open-circuit voltage and wire-feed speed was applied to the system, and the resulting current and arc voltage signals were measured. All experiments were run open-loop. Each experiment used the parameters given in Table 5.1. Table 5.2 shows the measured values of current and arc voltage with respect to open-circuit voltage and wire- feed speed as measured from the power supply sensors. The actual data acquired from the experiments was manipulated to obtain these results as follows. First, the transients in the output signals were neglected by deleting the first second of data. Then the remaining two seconds of data were averaged over time. The final values represent the steady-state values of current and arc voltage with respect to the constant values of open-circuit voltage and wire- feed speed.

Table 5.1: Welding Parameters Used During the Experiments

Weldment material	Flat stock mild steel
Weldment thickness	0.25"
Torch angle	0 degrees
Gas composition	85% Ar 15% CO_2
Gas pressure	30.0 psi
Electrode diameter	0.045"
Electrode composition	AWS/ASME SFA5.18
Initial X-axis speed	| 30 | ipm
Initial Y-axis speed	0.0 ipm
Power supply operating mode	Constant voltage
Open-circuit voltage operating range	28V to 38V
Wire-feed speed operating range	250 to 450 ipm
Contact tip-to-workpiece dist.	0.75"
Weld speed	30 ipm

5.2. EMPIRICAL MODELING OF A GMAW PROCESS

Table 5.2: Results of Steady-State Data Collection

Current	Arc Voltage	Open-circuit Voltage	Wire-Feed Speed
216.7478	34.4912	28	250
220.4587	35.7322	29	250
230.2010	37.1719	30	250
235.2805	38.4670	31	250
243.2863	39.7568	32	250
254.1635	40.8128	33	250
258.5118	41.9084	34	250
266.6107	43.1293	35	250
273.7057	44.3957	36	250
297.9041	45.1733	37	250
346.8698	45.0861	38	250
244.7135	33.9139	28	300
242.9675	35.1938	29	300
244.8064	36.5739	30	300
252.3317	38.0297	31	300
261.3481	39.4294	32	300
279.3352	40.4867	33	300
285.5294	41.6486	34	300
298.0920	42.8244	35	300
308.1061	43.8203	36	300
318.1792	44.7602	37	300
338.1852	45.5200	38	300
264.9320	33.5759	28	350
258.4770	35.1077	29	350
254.8920	36.6484	30	350
266.9976	37.5662	31	350
269.7978	35.8575	32	350
298.3253	40.0642	33	350
313.0345	41.5022	34	350
324.5637	42.7263	35	350
333.2261	43.7379	36	350
339.4062	44.6573	37	350
349.9693	45.2558	38	350
282.7422	33.1027	28	400
282.0588	34.5166	29	400
284.3064	35.7409	30	400
286.9252	36.7601	31	400
297.3010	38.1430	32	400
308.9067	39.7751	33	400
320.1427	41.3663	34	400
328.1787	42.6347	35	400
335.7934	43.5277	36	400
345.9308	44.3625	37	400
374.6583	45.0142	38	400
299.0219	33.0672	28	450
299.1571	34.2059	29	450
301.6897	35.5286	30	450
303.1589	36.6433	31	450
307.9232	37.8335	32	450
320.7758	39.2678	33	450
337.5388	40.4668	34	450
338.6300	42.4248	35	450
352.4701	43.1085	36	450
366.6525	43.9923	37	450
381.3725	44.6221	38	450

226 CHAPTER 5. CONTROL OF GMAW: A CASE STUDY

Using the data in Table 5.2, contour plots were created to show how the measured variables behave with respect to open-circuit voltage and wire-feed speed. Figure 5.2 shows the contour plot of average current with respect to open-circuit voltage and wire-feed speed. Note that current depends upon both open-circuit voltage and wire-feed speed. This implies that both open-circuit voltage and wire-feed speed can be manipulated to control current. Arc voltage is plotted in Figure 5.3

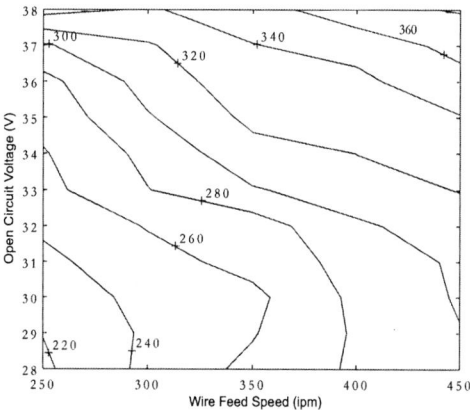

Figure 5.2: Experimental data - contour plot of average current (amps).

with respect to open-circuit voltage and wire-feed speed. Note that unlike current, arc voltage is heavily dependent upon open-circuit voltage and only slightly dependent upon wire-feed speed. Thus, manipulating only open-circuit voltage would enable a controller to achieve the desired arc voltage. Also, note that due to an error in measurement interpretation, at any given value of open-circuit voltage, the corresponding arc voltage reading at any wire-feed speed is always greater than the open-circuit voltage reading. From physics, and more specifically circuit theory, we know that this is not possible. Therefore, the measurement of arc voltage is incorrect. This effectively eliminates the possibility of optimizing the model for arc voltage. However, the general trends indicated in Figure 5.3 are correct, and future work will rectify this measurement error. Note that the trends indicated in Figures 5.2 and 5.3 are exploited below to develop multi-loop controllers for the GMAW process.

5.2. EMPIRICAL MODELING OF A GMAW PROCESS

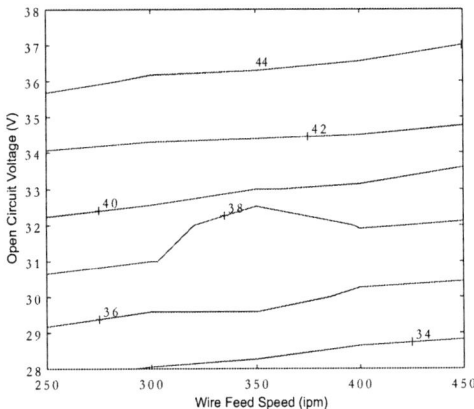

Figure 5.3: Data - contour plot of average arc voltage (volts).

Determination of Empirical Constants

A brute force method was used to optimize the fifth-order nonlinear mathematical model presented in Section 2.9. A program was written that simply ran the model for a given value of inputs and parameters, calculated the resulting steady-state current, and computed the associated error from the actual current at the particular operating point. This process was iterated to cover both the same range of inputs as the experiments described above and a wide range of parameters.

Analysis of the resulting data showed that the set of empirical parameters with the smallest summed squared error over the whole range of open-circuit voltage and wire-feed speed operating points was:

$R_a = 0.05$
$V_o = 5$
$E_a = 800$
$C_1 = 5 \times 10^{-10}$
$C_2 = 5 \times 10^{-11}$.

Figure 5.4 shows the resulting contour plot of current obtained from the model using these parameters. Notice that the plot looks similar to that of Figure 5.2. Both have hooks at smaller wire-feed speed and open-circuit voltage values and straighten out for higher values of the inputs. Figure 5.5 is a contour plot of the squared error (the error between Figure 5.2 and Figure 5.4). This represents the difference

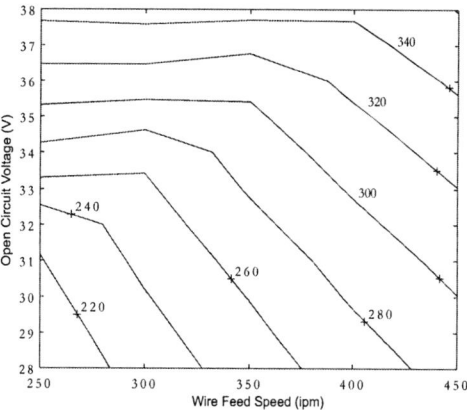

Figure 5.4: Simulation data - average current from the fifth-order model for the whole data set.

between the experimental data and the simulated data using the parameter set given above. Most of the error was encountered during high open-circuit voltage values and low values of wire-feed speed. An alternate approach to selecting the empirical parameters is to optimizing the model at each point in the range of the two inputs. That is, for each given operating point we find the best values for each of the empirical parameters. Then, when running the simulation, we use the values of the empirical parameters that correspond to the operating point of the simulation. Thus, we will have variable empirical "constants," where the variation is a function of the operating point. This produces a better fit, as seen in Figures 5.6 and 5.7. In contrast to Figure 5.4, which shows the simulated current contour for the case of "constant" constants, Figure 5.6, representing the simulated current contours using "variable" constants (an oxymoron to be sure!), is much closer to Figure 5.2, which shows the actual current contours. The difference is clear when comparing the error contours in Figure 5.5 and Figure 5.7.

To highlight the difference in the two optimization strategies, Figures 5.8 through 5.12 show contours of the parameters that would produce the smallest squared error of the average current at each point in the operating range of the inputs. Clearly, these "constants" are not con-

5.2. EMPIRICAL MODELING OF A GMAW PROCESS

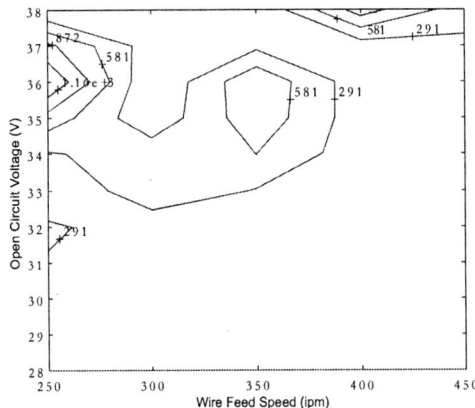

Figure 5.5: Simulation data - best squared error for the whole data set applications.

stants but rather depend on the operating condition. This means that different controller gains will be required at different operating points. Thus, the system is a good candidate for adaptive control.

230   CHAPTER 5.   CONTROL OF GMAW: A CASE STUDY

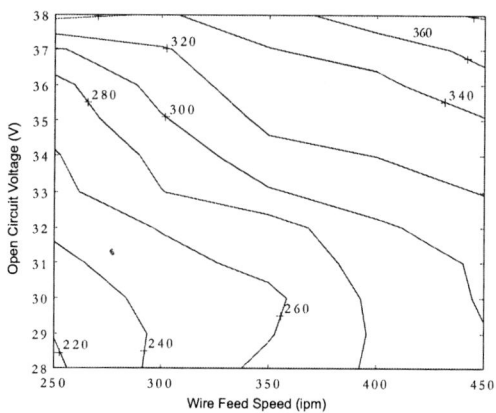

Figure 5.6: Simulation data - average current from the fifth-order model for individual data sets.

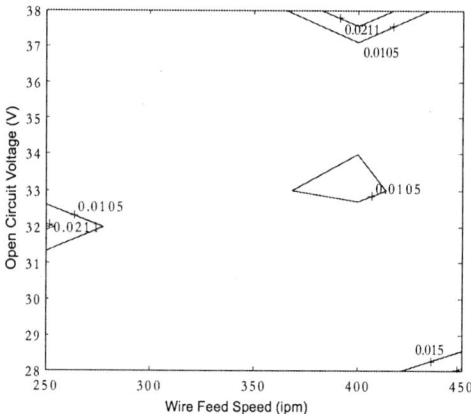

Figure 5.7: Simulation data - best squared error for individual data sets.

5.2. EMPIRICAL MODELING OF A GMAW PROCESS

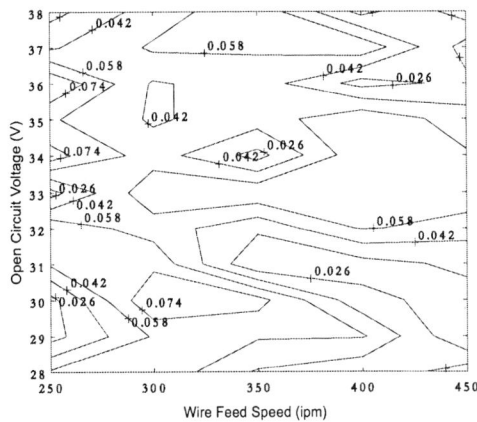

Figure 5.8: Simulation data - best R_a parameter for individual data sets.

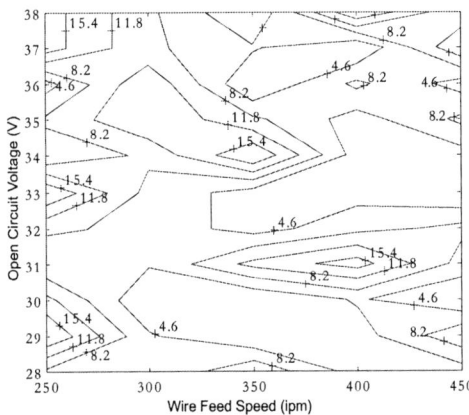

Figure 5.9: Simulation data - best V_o parameter for individual data sets.

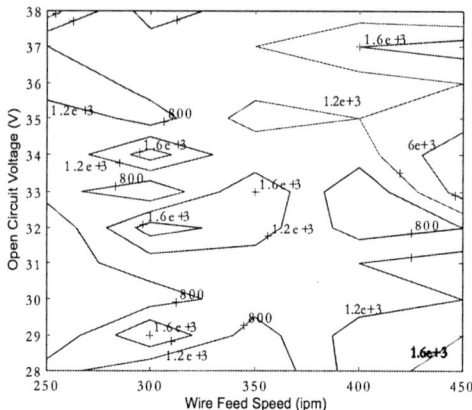

Figure 5.10: Simulation data - best E_a parameter for individual data sets.

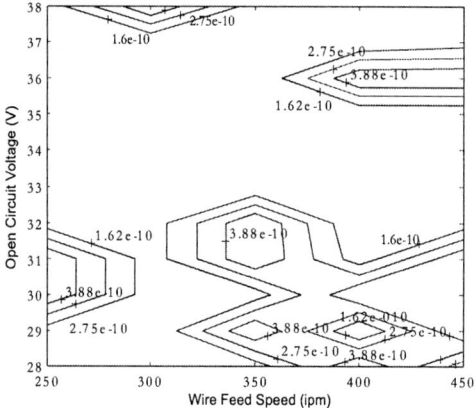

Figure 5.11: Simulation data - best C_1 parameter for individual data sets.

5.2. EMPIRICAL MODELING OF A GMAW PROCESS

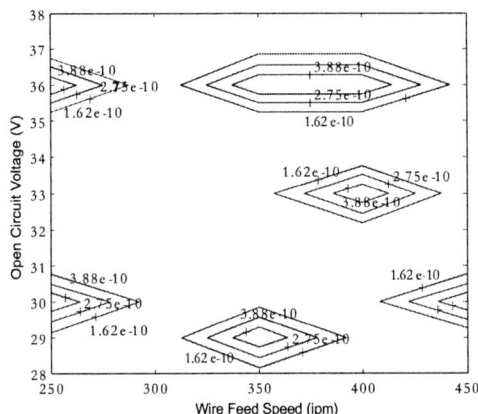

Figure 5.12: Simulation data - best C_2 parameter for individual data sets.

5.2.2 Empirical Transfer Function Model

In earlier work it was argued that it is reasonable to approximate the GMAW process with a two-by-two linear, multivariable system (Yender 1997, Moore et al. 1997a). On this assumption, from simple step response tests it is possible to derive an empirical model in transfer function form. Figures 5.13 through 5.16 show graphs of the conditioned measurements resulting from step changes in both the wire-feed speed and open-circuit voltage. Included on the graphs are the

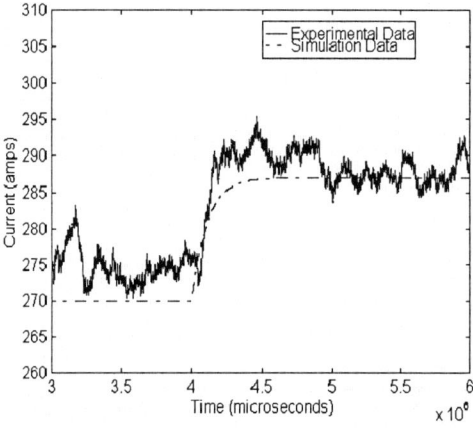

Figure 5.13: Current response to a step increase in the wire-feed speed.

corresponding empirical models fit to the conditioned response. These models were developed heuristically using MATLAB® simulations. Figure 5.13 shows the rise in current due to a step increase in the wire-feed speed from 350 ipm to 400 ipm. Figure 5.16 shows a drop in the measured voltage for the same step in wire-feed speed. Figure 5.14 shows a rise in the measured voltage for a step increase in open-circuit voltage from 28 volts to 32 volts. Figure 5.15 shows the response of the current measurement due to the same step in open-circuit voltage.

A model derived from the step responses in Figures 5.13 through 5.15, collected in a transfer matrix form, is given as

$$\begin{bmatrix} I(s) \\ V_{arc}(s) \end{bmatrix} = \begin{bmatrix} \dfrac{0.34}{0.11s+1} & \dfrac{68800}{s^2+42.7s+6400} \\ \dfrac{-8.6 \times 10^{-3}}{0.034s+1} & \dfrac{1.35}{0.083s+1} \end{bmatrix} \begin{bmatrix} S(s) \\ V_{oc}(s) \end{bmatrix}$$

5.3. SISO CURRENT CONTROL USING PI CONTROLLER 235

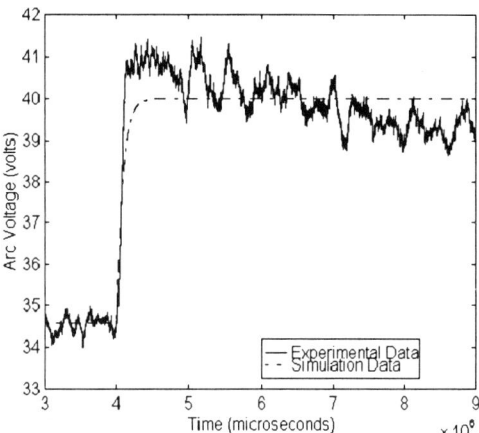

Figure 5.14: Voltage response to a step increase in the open-circuit voltage.

where I is current, V_{arc} is arc voltage (measured from the power supply), S is wire-feed speed, and V_{oc} is open-circuit voltage.

5.3 SISO Current Control Using PI Controller

In this section, the experimental results obtained using a PI controller for the control of single-input, single-output (SISO) current will be presented. It was observed from the modeling results of the previous section that the step change in open-circuit voltage has more effect on the current than a step change in the wire-feed speed. Therefore, for the control of current, open-circuit voltage was chosen to be the control input.

The continuous PI controller transfer function is given in general form as follows:

$$G_c(s) = \frac{U(s)}{E(s)} = K_p + K_i \frac{1}{s} \quad (5.1)$$

where K_p and K_i are the proportional and integral gains, respectively. $U(s)$ is the control signal and $E(s)$ is the error signal. Since the controller is implemented digitally, a discrete PI controller is used that takes the following form:

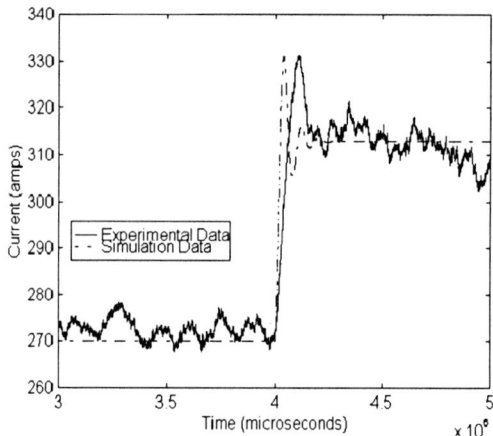

Figure 5.15: Current response to a step increase in the open-circuit voltage.

Proportional Action:
$$P(k) = K_p e(k) \qquad (5.2)$$

Integral Action:
$$I(k) = I(k-1) + K_i T_s e(k) \qquad (5.3)$$

PI Action:
$$u(k) = P(k) + I(k) = K_p e(k) + I(k-1) + K_i T_s e(k) \qquad (5.4)$$

where $e(k)$ and T_s are the error signal and sampling time, respectively. The block diagram of the closed-loop system is depicted in Figure 5.17. Welding parameters were as defined in Table 5.3. In addition, during the experiments, weld speed and wire-feed speed were set to 30 ipm and 350 ipm, respectively. In each experiment, the controller was turned on at 3 sec. In other words, the process was run open-loop for the first 3 sec. During the experiments, the initial value of the open-circuit voltage V_{oc} was set to 28V. Total weld time was 7 sec. for each experiment.

Numerous experiments were conducted. A representative result is shown in Figure 5.18, which gives the current output for a series of step changes. In the experiment, the system ran open loop for 3 seconds

5.3. SISO CURRENT CONTROL USING PI CONTROLLER

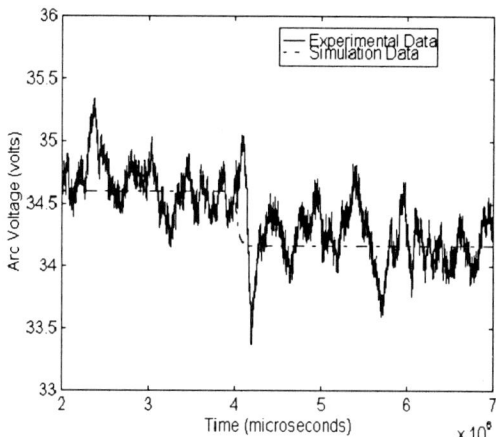

Figure 5.16: Voltage response to a step increase in the wire-feed speed.

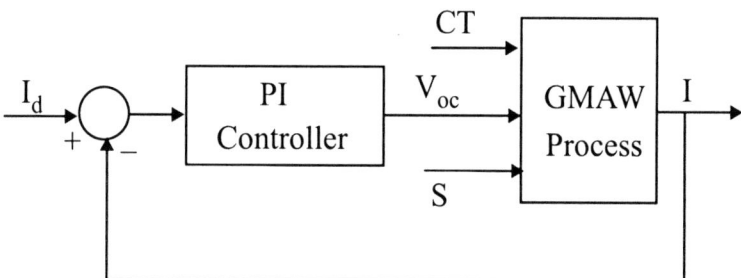

Figure 5.17: Closed-loop system with PI controller.

with $V_{oc} = 28V$, then the controller is turned on with a set point of 260A between 3 and 5 seconds and a set point of 240A after that time. Note that the effect of the controller, as designed, is to change the system response from the nominally underdamped characteristic of the open-loop system (see Figure 5.15) to a closed-loop system response that is overdamped. As can be seen, the system tracks the set points quite well. It is also important to note the associated control input voltage and measured arc voltage, shown in Figures 5.19 and 5.20, respectively. Clearly, because we are only controlling the current, the arc voltage varies as a result of our controller action.

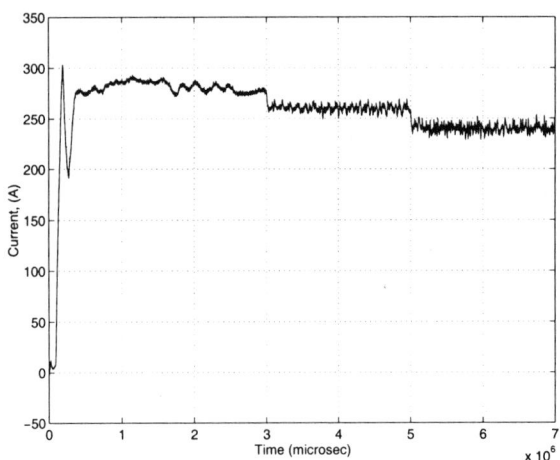

Figure 5.18: Current response, $K_p = 0.5$, $K_i = 5$, desired current=260A, actual current=260.0033A for $3 > t \leq 5$ and desired current=240A, actual current=240.0955A for $5 > t \leq 7$.

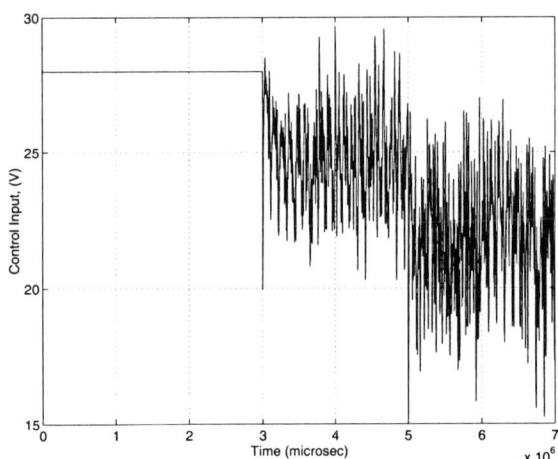

Figure 5.19: Control signal, open-circuit voltage, for $K_p = 0.5$, $K_i = 5$, desired current=260A for $3 > t \leq 5$ and desired current=240A for $5 > t \leq 7$.

5.4. MULTI-LOOP CONTROL OF THE GMAW PROCESS

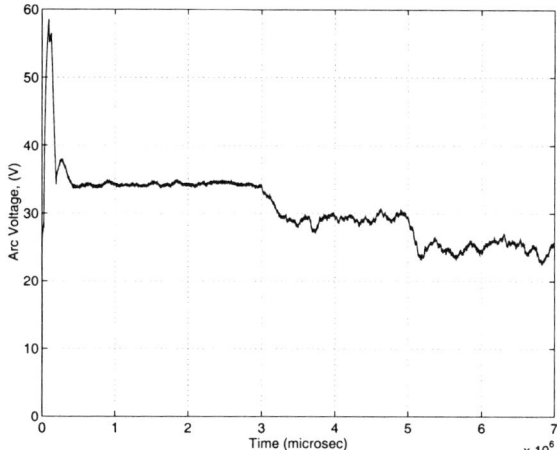

Figure 5.20: Arc voltage, for $K_p = 0.5$, $K_i = 5$, desired current=260A for $3 > t \leq 5$ and desired current=240A for $5 > t \leq 7$.

5.4 Multi-Loop Control of the GMAW Process

Although the previous section showed that single-loop control of current can be effective, it may be desirable to control both current and voltage to desired set points (this will eventually allow us to independently control mass and heat transfer in the process). In this case one may use a multivariable or a multi-loop strategy. In [754] and [755] and later in this chapter, results are presented using adaptive multivariable control. In this section results from multi-loop control experiments are presented. For a multi-loop control structure the dominant pairing must be determined. At the outset, based on the modeling results in Section 3, because the wire-feed speed is observed to have such a little effect on measured voltage, it might be suspected that the correct pairings are to use wire-feed speed to control current and open-circuit voltage to control arc (measured) voltage. As it turns out, the relative gain array analysis agrees with this pairing. Relative gain array (RGA) analysis is an analytical approach to choosing loop pairings that is popular in the process control community [756].

5.4.1 Relative Gain Array Analysis

The first step to compute the array is to assemble the steady-state gains from the system transfer matrix in the following steady-state gain matrix, K:

$$K = \begin{bmatrix} 0.34 & 10.75 \\ -8.6 \times 10^{-3} & 1.35 \end{bmatrix}$$

where the first column corresponds to wire-feed speed, the second column corresponds to open circuit voltage, the first row corresponds to current, and the second row corresponds to arc (measured) voltage. The next step is to determine the inverse transpose of K

$$K^{-1} = \begin{bmatrix} 2.4481 & 0.0156 \\ -19.4941 & 0.6166 \end{bmatrix}$$

Then the two matrices are multiplied element by element to arrive at the RGA

$$R = \begin{bmatrix} 0.8324 & 0.1676 \\ 0.1676 & 0.8324 \end{bmatrix}$$

Since the values closest to unity appear on the main diagonal, the RGA method indicates that wire-feed speed should be paired with current and open-circuit voltage should be paired with arc voltage [756]. This implies the architecture shown in Figure 5.21.

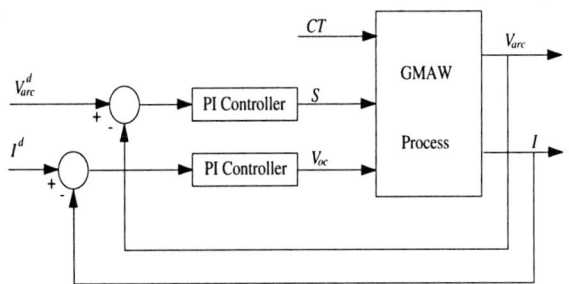

Figure 5.21: Multi-loop control of the GMAW process.

5.4.2 Multi-loop Control Experimental Results

Once the pairings for inputs and outputs have been found, design of a controller can begin. Initially, selection of a control algorithm must be

5.4. MULTI-LOOP CONTROL OF THE GMAW PROCESS

done, and then analysis of the controller's effect on the system must be determined. Due to its simplicity to implement and its ability to reject steady-state step input, a PI control algorithm is selected to control each paired input and output. Because two PI controllers will be used, there are four controller gains that must be tuned based on the desired performance criteria of the outputs. Several methodologies exist to tune all of the controller gains at once. However, because the wire-feed speed has such a little effect on the arc voltage, initial attempts to tune the PI controller gains are constrained to analysis involving only the transfer function describing the change in arc voltage due to a change in open circuit voltage. This analysis is identical to analysis that would be done in a SISO case where arc voltage is controlled by open-circuit voltage. Once the PI controller gains were determined for the simulated SISO case, they will not be changed, and a PI controller on the wire-feed speed-current pair will be added. The gains for the additional PI controller were then determined by trial and error analysis based on simulated results utilizing all four transfer functions. In the transfer function domain, the resulting PI control structure had the form

$$V_{oc}(s) = \frac{0.25s + 5}{s}(V_{arc_d} - V_{arc})$$

and

$$S(s) = \frac{0.01s + 5}{s}(I_d - I)$$

As in the SISO case above, in the experiments, these controllers were actually implemented with a discrete-time algorithm.

A variety of experiments were conducted with the multi-loop controller. Changes in open-circuit voltage and wire-feed speed were made about the nominal operating point of 28 volts and 350 ipm (the operating point corresponds to the operating point from which empirical model identification was done). Figures 5.22 and 5.23 show typical experimental results versus simulation results (in the interest of space the manipulated variables, V_{oc} and S, are not shown). As shown in Figure 5.22, there is agreement between the experimental and simulated current, especially in the steady-state tracking property, although the transient response is not so close. Figure 5.23 also shows considerable agreement between simulated and experimental results. It is clear that the PI multi-loop control algorithm provides for set point tracking.

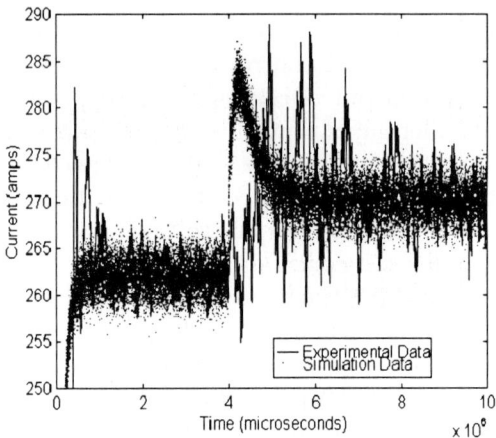

Figure 5.22: Experimental results: current response, with simulation data.

5.4.3 Disturbance Rejection Test

A key feature of a control system is its ability to reject disturbances. Processes are often subjected to extraneous signals that influence the output. The controller should be able to adjust the system inputs to counteract the effect of such signals. In the GMAW process, one disturbance that can occur is an unexpected change in the contact tip-to-workpiece distance (CT). In an open-loop setting, one expects such a disturbance to cause an increase in the measured arc voltage. An experiment was conducted to test the ability of the multi-loop control system to maintain desired set points of current and arc voltage in the face of such a disturbance. The disturbance was simulated by welding over the edge of a thin plate tack-welded onto a base plate. The thin plate on which the weld was started was 1/8 inch thick so the disturbance appeared as a "step" dip in the metal of an eighth of an inch when the torch head moved over the edge of the thin plate.

First, an open-loop test was conducted. The controllable inputs (i.e., wire-feed speed and open-circuit voltage) were held constant over the disturbance at values of 340 ipm and 30.5 volts, respectively. The results are shown in Figures 5.24 and 5.25. From Figure 5.24, it appears that the average current doesn't seem to change much due to the disturbance (which occurred at 4 seconds). However, there does appear

5.4. MULTI-LOOP CONTROL OF THE GMAW PROCESS 243

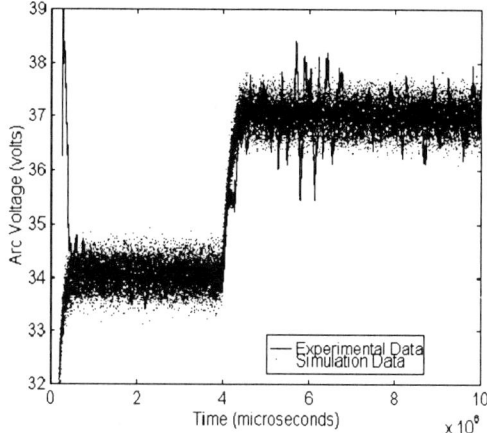

Figure 5.23: Experimental results: voltage response, with simulation data.

to be some influence on the variance of the current. The disturbance seemed to cause greater variability in the current response. Figure 5.25 shows that the disturbance has more of an effect on the arc voltage. It appears that the voltage settles at 34 volts with little variance before the disturbance. Following the disturbance a somewhat large spike in voltage occurred, and after about a second the voltage rose to a value near 36 volts with more variance. Next, the experiment was repeated with the multi-loop controller in place. Figures 5.26 through 5.29 show the resulting signals. The CT disturbance occurs at 4 seconds. Figure 5.26 shows that (for a set point value of 270 amps) the current is maintained at 270 amps. Note that the variability of the current is less affected by the CT disturbance than in the open-loop situation. Figure 5.27 shows the commanded wire-feed speed signal that ensured that the current be maintained at 270 amps. Figure 5.28 shows that (for a set point value of 34 volts) the arc voltage is maintained at 34 volts. Again, the variability of the arc voltage is much less than that in Figure 5.25. Figure 5.29 shows the commanded open-circuit voltage signal that ensured that the arc voltage be maintained at 34 volts.

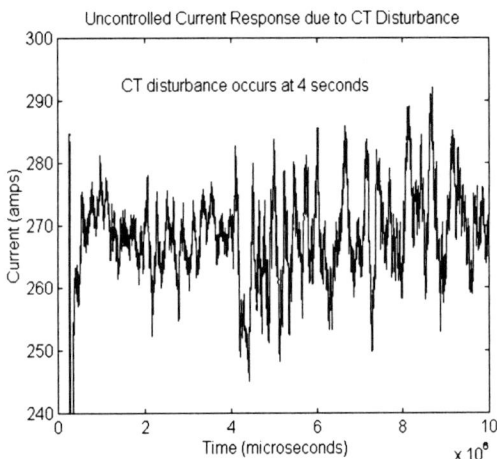

Figure 5.24: Experimental results: open-loop current response, step disturbance.

5.5 Adaptive Control of GMAW Process

In this section we present the design and implementation of a Direct Model Reference Adaptive Control (DMRAC) for a gas metal arc welding (GMAW) process. Two models are used to control the system. First, a highly nonlinear fifth-order mathematical model for the GMAW process is used to describe the system dynamics. This model is the basis for the process-level control of current and arc voltage to desired set points. Second, a model of the heat and mass transfer from the process to the weld pool is presented. This model is used as the basis for prescribing the desired set points to be used by the process-level controller. Specifically, an optimization technique is used to find an input vector (contact tip-to-workpiece distance, open-circuit voltage, wire-feed speed and weld speed) and a corresponding output vector (physically possible current and stick-out) that result in maximum production rate for prescribed values of heat and mass transfer. For process-level control, the GMAW process is modeled as two-input, two-output second-order system of differential equations. Current and arc voltage are the process outputs, which are controlled by open-circuit voltage and wire-feed speed, which are chosen as the process control inputs. As we have noted, because of the nonlinear nature of the process, the system transfer function varies as a function of operating point. For

5.5. ADAPTIVE CONTROL OF GMAW PROCESS

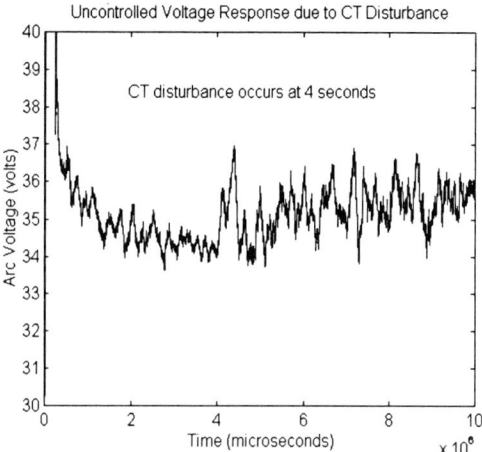

Figure 5.25: Experimental results: open-loop voltage response, step disturbance.

this reason, a DMRAC is designed and then implemented on the actual GMAW process. Satisfactory experimental results are obtained. Hence, this new direct adaptive control algorithm allows the control of both current and arc voltage and consequently makes it possible to achieve high quality weldings in GMAW processes.

5.5.1 Overview

Although industrial practices for GMAW are well-established, in many situations, automatic control of the process is highly desirable in order to produce improved quality in weld properties. However, the issue of optimal strategies and architectures for automated GMAW control remains open. The importance of this problem is self-evident. High-quality welding procedures are essential to overall product quality in any industrial production setting. Through the use of automatic control, it may be possible to produce more consistent weld quality as well as to eliminate much of the "guesswork" involved in setting power supply parameters to the proper value required to achieve good welds. Advanced methods for controlling arc welding power supplies can lead to significant improvements in the economic competitiveness of industry.

Other researchers have considered the problem of GMAW control.

246   CHAPTER 5. CONTROL OF GMAW: A CASE STUDY

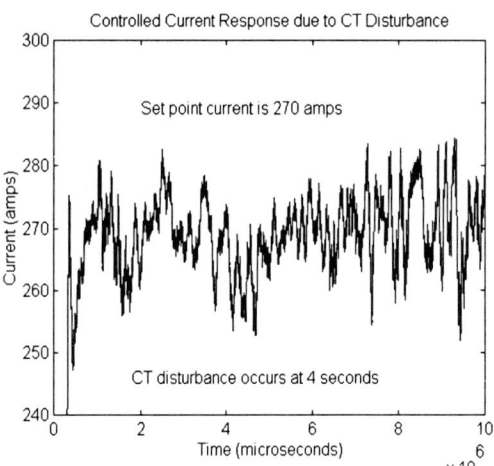

Figure 5.26: Experimental results: closed-loop current response, disturbance rejection.

In the area of classical control, Smartt and Einerson [757] developed a steady-state model for heat (given on a per unit length of weld) and mass (the transverse cross-sectional area of the deposited metal) transferred from the electrode to the workpiece in a GMAW process. Using the relations for heat and mass, they developed a PI based control system for maintaining the desired heat and mass by regulating the current. In what follows we will use this same model to determine current and voltage set point for our DMRAC.

In the previous section, we have presented a classical approach to controlling the GMAW process. In the area of adaptive control, a successful application of a pseudo-gradient adaptive algorithm for self-tuning a PI-based puddle width controller for a consumable electrode GMAW process was given by Henderson et al. [758, 759]. Doumanidis also developed an adaptive multi-input multi-output (MIMO) scheme to control both geometrical and thermal characteristics of a weld based on lumped parameter and distributed parameter modeling and identification [760]. The problem of adaptive and decoupling control of the MIMO welding process was also addressed by Cook et al. [761]. Other control methodologies have also been developed, including the feedback linearization technique that was applied to the fifth-order nonlinear model described below [762, 763]. Here, a simplified second order

5.5. ADAPTIVE CONTROL OF GMAW PROCESS

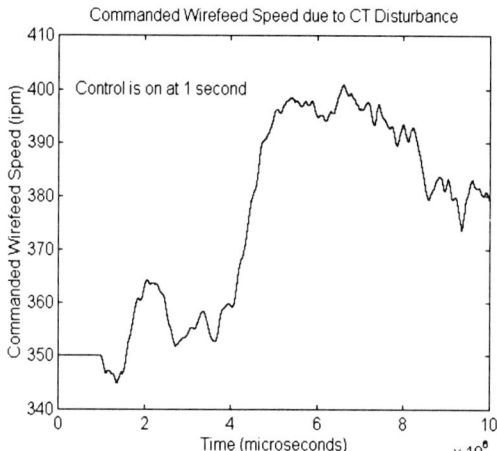

Figure 5.27: Experimental results: closed-loop wire-feed speed response, disturbance rejection.

nonlinear model and the models for heat and mass transfer were used to independently control current and arc length using the open-circuit voltage and wire-feed speed.

In this section, we consider a comprehensive approach to the problem of controlling the GMAW process. Since the mass and heat transferred to the workpiece determine the quality of welding, it is necessary to achieve the desired mass and heat values for a particular welding. Assuming contact tip-to-workpiece distance and weld speed are held constant during welding, one approach to controlling the quality of the weld in the GMAW process is to control the current and arc voltage so as to control the heat and mass input to the weld pool. Our approach to achieving this is to first use the Smartt/Einerson model of the heat and mass transfer from the process to the weld pool [757] as the basis for prescribing the desired set points to be used by a process-level controller that regulates current and arc voltage. Then, a process-level controller actually regulates the current and arc voltage to their prescribed set points. To determine these set points from the Smartt/Einerson model, an optimization technique is used to find an input vector (contact tip-to-workpiece distance, open-circuit voltage, wire-feed speed and weld speed) and a corresponding output vector (physically possible current and stick-out, or arc voltage) that result in maximum production rate for prescribed values of heat and mass transfer. The resulting input and output vectors then serve to define nominal operating conditions. This method maximizes the weld speed with respect to current, contact tip to work piece distance, stick-out, and weld speed for any given desired mass and heat values. Constraints are the welding parameters, which vary within their upper and lower bounds and the equilibrium condition of the process at steady-state. A feasible region in the plane defined by the mass and heat inputs has been determined. This development allows the high quality welding to be performed at the maximum production rate and increases the productivity in GMAW processes.

For the process-level control, the GMAW process is modeled as two-input two-output second-order system of differential equations. Current and arc voltage are the process outputs, which are controlled by open-circuit voltage and wire-feed speed, which are chosen as the process control inputs. Because of the nonlinear nature of the process, the system transfer function varies as a function of operating point. Thus, classical techniques as described above may not be adequate. Therefore, the design of a MIMO DMRAC for a GMAW process is considered. Attractive features of this type of adaptive control include lack of dependence on process parameter estimates, control calculation which does not require adaptive observers or full state feedback, applicability to MIMO plants, and ease of implementation [764]. However, asymptotic stability is ensured provided that the plant is almost strictly positive real (ASPR). That is, there exists a feedback gain (not needed for implementation) such that the resulting closed-loop transfer function matrix is strictly positive real (SPR) [764].

5.5.2 Model Simplification and Linearization

Consider again the dynamics of the process given by the following state-space equations:

$x_1 = x$: droplet displacement (m)
$x_2 = \dot{x}$: droplet velocity (m/sec)
$x_3 = m_d$: droplet mass (kg)
$x_4 = l_s$: stick-out (m)
$x_5 = I$: current (A)

5.5. ADAPTIVE CONTROL OF GMAW PROCESS

$$\begin{aligned}
\dot{x}_1 &= x_2 \\
\dot{x}_2 &= \frac{-Kx_1 - Bx_2 + F_{tot}}{x_3} \\
\dot{x}_3 &= M_R \rho_w \\
\dot{x}_4 &= u_1 - \frac{M_R}{\pi r_w^2} \\
\dot{x}_5 &= \frac{u_2 - (R_a + R_s + R_L)x_5 - V_0 - E_a(CT - x_4)}{L_s}
\end{aligned} \quad (5.5)$$

The dynamics of the GMAW process given by (5.6) are highly *nonlinear* and quite complex. This makes the application of the modern control strategies difficult. Therefore, in the following a simplified model is presented based on valid approximations. Consider first the current I and stick-out l_s relations from (5.6)

$$\begin{aligned}
\dot{x}_4 &= u_1 - \frac{M_R}{\pi r_w^2} \\
\dot{x}_5 &= \frac{u_2 - (R_a + R_s + R_L)x_5 - V_0 - E_a(CT - x_4)}{L_s}
\end{aligned} \quad (5.6)$$

It is obvious that the stick-out distance ($l_s = x_4$) is much larger than the sum of the droplet radius (r_d) and the drop distance (x_1) or mathematically

$$x_4 \gg \frac{1}{2}\left(\left(\frac{3x_3}{4\pi \rho_w}\right)^{\frac{1}{3}} + x_1\right) \quad (5.7)$$

This approximation simplifies the above equation (5.6) yielding

$$\begin{bmatrix} \dot{x}_4 \\ \dot{x}_5 \end{bmatrix} = \underbrace{\begin{bmatrix} 0 & -\frac{C_1}{\pi r_w^2} \\ \frac{E_a}{L_s} & -\frac{R_a + R_s}{L_s} \end{bmatrix}}_{A} \underbrace{\begin{bmatrix} x_4 \\ x_5 \end{bmatrix}}_{X} +$$

$$\underbrace{\begin{bmatrix} -\dfrac{C_2\rho}{\pi r_w^2} x_4 x_5^2 \\ -\dfrac{1}{L_s}(\rho x_4 x_5 + V_0 + E_a CT) \end{bmatrix}}_{f(X,CT)} + \underbrace{\begin{bmatrix} 1 & 0 \\ 0 & \dfrac{1}{L_s} \end{bmatrix}}_{B} \underbrace{\begin{bmatrix} u_1 \\ u_2 \end{bmatrix}}_{u} \tag{5.8}$$

$$\begin{bmatrix} \delta y_1 \\ \delta y_2 \end{bmatrix} = \underbrace{\begin{bmatrix} -E_a & R_a \\ 0 & 1 \end{bmatrix}}_{C_p} \begin{bmatrix} \delta x_4 \\ \delta x_5 \end{bmatrix} \tag{5.9}$$

The relation (5.8) shows that the dominant states, current and stickout are independent of the other states. This allows us to analyze the system using only two equations (5.8) rather than five equations (5.6). Then linearizing the system about steady-state yields

$$\begin{bmatrix} \delta \dot{x}_4 \\ \delta \dot{x}_5 \end{bmatrix} = \underbrace{\begin{bmatrix} -\dfrac{C_2 \rho \bar{x}_5^2}{\pi r_w^2} & -\dfrac{(C_1 + 2C_2 \rho \bar{x}_4 \bar{x}_5)}{\pi r_w^2} \\ \dfrac{E_a - \rho \bar{x}_5}{L_s} & -\dfrac{(R_a + R_s + \rho \bar{x}_4)}{L_s} \end{bmatrix}}_{A_p} \begin{bmatrix} \delta x_4 \\ \delta x_5 \end{bmatrix} + \underbrace{\begin{bmatrix} 1 & 0 \\ 0 & \dfrac{1}{L_s} \end{bmatrix}}_{B_p} \begin{bmatrix} \delta u_1 \\ \delta u_2 \end{bmatrix} \tag{5.10}$$

$$\begin{bmatrix} \delta y_1 \\ \delta y_2 \end{bmatrix} = \underbrace{\begin{bmatrix} -E_a & R_a \\ 0 & 1 \end{bmatrix}}_{C_p} \begin{bmatrix} \delta x_4 \\ \delta x_5 \end{bmatrix} \tag{5.11}$$

where $\bar{x}_4$ and $\bar{x}_5$ correspond to the steady-state values of x_4 and x_5, respectively, C_1, C_2 are the melting rate constants, and ρ is the resistivity of the electrode. This linearized model (5.11) will be used below.

5.5. ADAPTIVE CONTROL OF GMAW PROCESS

5.5.3 Model for Heat and Mass Transfer

This part is heavily dependent upon the work of Smartt and Einerson [757]. A steady-state model of the GMAW process is obtained for electrode melting and heat and mass transfer from the electrode to the workpiece.

The electric power consumed by the process is approximately equal to the sum of that consumed by the resistive heating of the electrode and that consumed by the arc as

$$IE = IV_e + IV_{arc} \qquad (5.12)$$

where V_e is the voltage drop across the electrode and V_{arc} is the voltage drop across the arc. The heat input H to the base metal per unit length of weld is given by

$$H = \frac{EI\eta}{R} = \frac{I(V_e + V_{arc})\eta}{R} \qquad (5.13)$$

where E is the secondary circuit voltage drop, η is the heat transfer efficiency from the process to the base metal, and R is the weld speed. The weld reinforcement G, defined as the transverse cross-sectional area of the deposited metal, is given as

$$G = \frac{AS}{R} = \frac{M_R}{R} \qquad (5.14)$$

where A is the cross-sectional area of the electrode, S is the electrode-feed speed, M_R is the melting rate, and R is the weld speed.

Since the mass G and heat H transferred to the workpiece determine the quality of welding, it is necessary to investigate the relation between $G - H$ and the welding parameters. In other words, to determine what the welding parameters are for given desired $G - H$ values.

5.5.4 Feasibility Region in the G − H Plane

Our objective is to determine all values of the desired $G - H$ at which there exist an input vector (CT, V_{oc}, S, R) and a corresponding output vector (I, l_s) that are physically possible. In other words, we need to determine a *feasible* region in the $G - H$ plane. One way of finding this *feasible* region is to perform the following optimization method [765], where underscores and overbars are used to denote limiting values of a variable.

$$\text{maximize} \ \{R\}_{R,\ CT,\ I,\ l_s} \qquad (5.15)$$

$$\text{subject to}: \begin{cases} \underline{CT} \leq CT \leq \overline{CT} \\ \underline{f(I)} \leq V_{oc} \leq \overline{f(I)} \\ \underline{S} \leq S \leq \overline{S} \\ \underline{I} \leq I \leq \overline{I} \\ \underline{l_s} \leq l_s \leq CT \\ \underline{R} \leq R \leq \overline{R} \\ \overline{u} = -B^{-1}(A\overline{X} + f(\overline{X},\overline{CT})) \end{cases} \qquad (5.16)$$

We should note that the weld speed R is chosen as the objective function so that maximum productivity is achieved. In the above optimization, all the constraints except the last one are physical constraints. The last one corresponds to the dynamics of the system and is obtained from the steady-state condition. Equation (5.8) must also be satisfied at steady-state. That is why we need to include it as a constraint in the above optimization.

Performing the optimization method given by (5.15), we obtained the feasible values for $G - H$ satisfying the constraint set given by (5.16). The ranges considered for G, H, and the weld parameters are as follows:

$$\begin{aligned} 1.5e-5 &\leq G \ [m^2] &&\leq 7e-5 \\ 0.5e6 &\leq H \ [J/m] &&\leq 2.5e6 \\ 0 &< CT \ [m] &&\leq 25e-3 \\ 0.021 &\leq S \ [m/s] &&\leq 0.33 \\ 25 &\leq I \ [A] &&\leq 565 \\ 11.51 - 0.0204I &\leq V_{oc} \ [V] &&\leq 50.6 - 0.0241I \\ 1e-3 &\leq l_s \ [m] &&\leq CT \\ 1e-3 &\leq R \ [m/s] &&\leq 25e-3 \end{aligned} \qquad (5.17)$$

The feasible region is shown Figure 5.30. Weld speed and current changes with respect to $G - H$ values are shown in Figures 5.31 and 5.32, respectively. That is, Figure 5.31 shows the optimal R that can be achieved for given values of G and H and Figure 5.32 shows the corresponding values of current.

5.5. ADAPTIVE CONTROL OF GMAW PROCESS

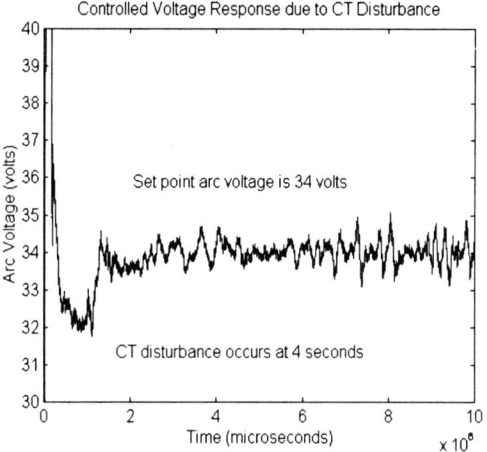

Figure 5.28: Experimental results: closed-loop voltage response, disturbance rejection.

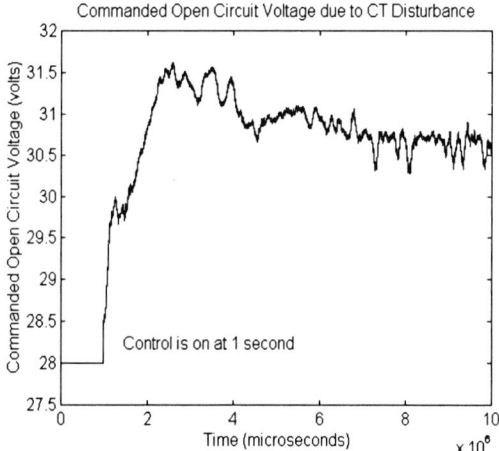

Figure 5.29: Experimental results: open-loop open-circuit voltage response, disturbance rejection.

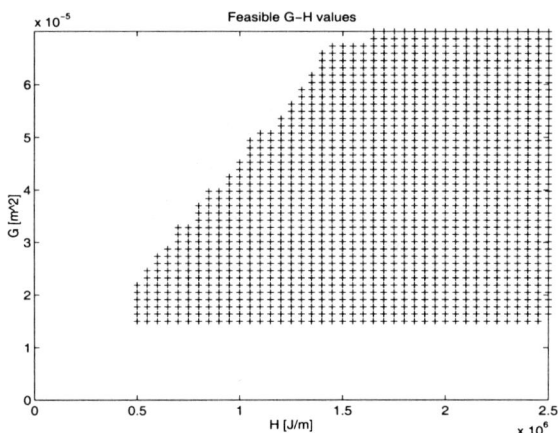

Figure 5.30: Feasible region in the $G - H$ plane.

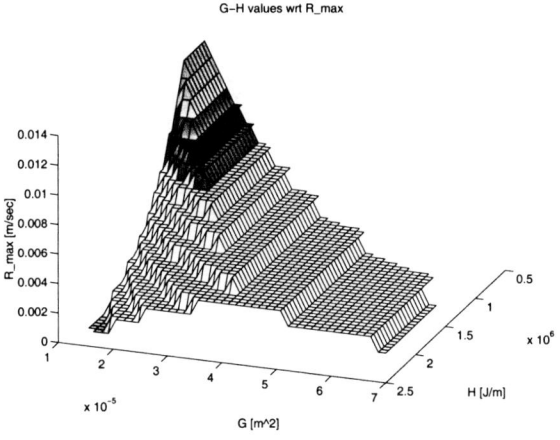

Figure 5.31: Weld speed with respect to $G - H$ values.

5.5. ADAPTIVE CONTROL OF GMAW PROCESS

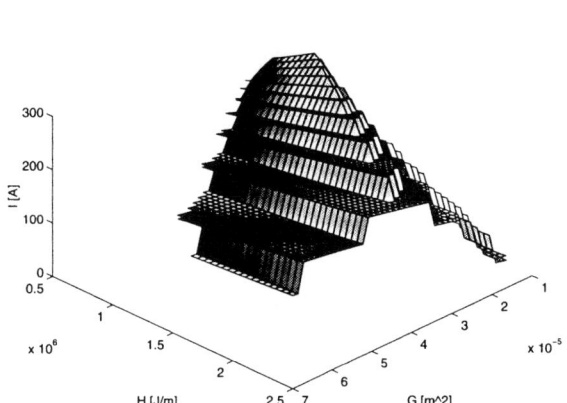

Figure 5.32: Current changes in the $G - H$ plane.

5.5.5 Stability Analysis and Characterization of G − H Plane

It is necessary to have some information about the stability of the system. In the feasible region, we need to know if there is any region that corresponds to an unstable system. On the other hand, if the system is stable, then we need to know whether it is overdamped or underdamped for the given desired $G - H$ values. In order to answer these questions, consider first the linearized second-order system given by (5.10). To determine whether the system represented by (5.10) is stable or not, it is necessary to find the eigenvalues of the matrix A or the roots of the characteristic equation of the matrix A. The characteristic equation is obtained from

$$| sI - A | = 0 \tag{5.18}$$

where I is the identity matrix. Equation (5.18) yields to the following characteristic polynomial

$$s^2 + a_1 s + a_0 = 0 \tag{5.19}$$

where

$$a_1 = [R_s \pi r_w^2 + R_a \pi r_w^2 + L_s C_2 \rho \bar{x}_5^2 + \rho \bar{x}_4 \pi r_w^2]/[L_s \pi r_w^2]$$
$$a_0 = [E_a C_1 + R_s C_2 \rho \bar{x}_5^2 - \rho \bar{x}_5 C_1 + R_a C_2 \rho \bar{x}_5^2 + 2 E_a C_2 \rho \bar{x}_5 \bar{x}_4 -$$
$$\rho^2 \bar{x}_4 C_2 \bar{x}_5^2]/[L_s \pi r_w^2] \tag{5.20}$$

It is not straight forward to check the stability of this characteristic polynomial, since the welding parameters $\bar{x}_4$ and $\bar{x}_5$ can vary in a prescribed range as given by (5.17). These type of polynomials are often called *interval* polynomials. A stability check for interval polynomials requires robust stability analysis such as Kharitonov's theorem [766]. Therefore, we applied Kharitonov's theorem to the above characteristic polynomial (5.19) and found that it is stable for the given range of parameters. It means that the linearized system in the entire feasible region is stable. Having determined that the system is stable, we need next to investigate whether the system is underdamped or overdamped and to characterize the feasible region in terms of the system's response. Analysis of the roots of the characteristic polynomial shows that the linearized system is underdamped in almost the entire feasible region. This is shown in Figure 5.33. We found that the damping ratio of the

5.6. CONTROL STRATEGY

system is bounded by

$$0.69 \leq \zeta \tag{5.21}$$

Figure 5.34 shows the distribution of the eigenvalues of the system in the $G - H$ plane.

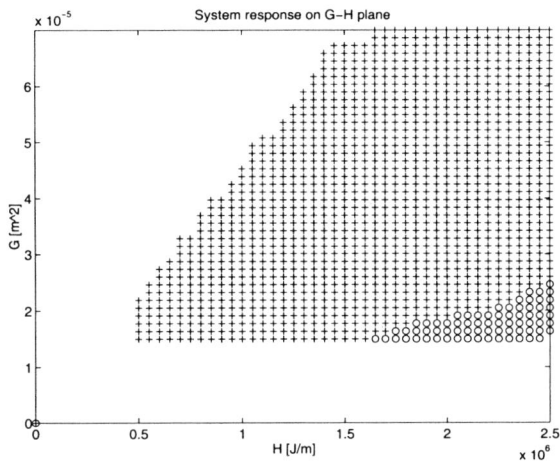

Figure 5.33: Distribution of the system response in the $G - H$ plane – (+): underdamped, (o): overdamped.

5.6 Control Strategy

Having modeled the GMAW process in the previous section, defined its inputs and outputs, established nominal operating conditions, and discussed the open-loop system's dynamic characteristics (stability and poles), let us consider control of the process. Although this system has three inputs and two outputs, we will assume that the contact tip-to-work piece distance CT is constant for a particular welding process and model the system with two inputs and two outputs. In this section, implementation of the direct model reference adaptive control will be discussed.

5.6.1 Formulation of the DMRAC

We begin by summarizing the necessary theory. The linear time invariant model reference adaptive control problem is considered for a plant

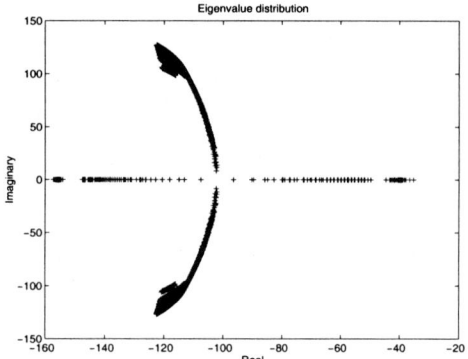

Figure 5.34: Eigenvalue distribution in the $G - H$ plane.

given by [764]

$$\begin{aligned} \dot{x}_p(t) &= A_p x_p(t) + B_p u_p(t) \\ y_p(t) &= C_p x_p(t) \end{aligned} \quad (5.22)$$

where $x_p(t)$ is the $(n \times 1)$ state vector, $u_p(t)$ is the $(m \times 1)$ control vector, $y_p(t)$ is the $(q \times 1)$ plant output vector, and A_p, B_p, and C_p are matrices with appropriate dimensions. The range of the plant parameters is assumed to be bounded as defined by

$$\begin{aligned} \underline{a}_{ij} \leq a_p(i,j) \leq \overline{a}_{ij}, i,j = 1, \cdots, n \\ \underline{b}_{ij} \leq b_p(i,j) \leq \overline{b}_{ij}, i,j = 1, \cdots, n \end{aligned} \quad (5.23)$$

where $a_p(i,j)$ is the $(i,j)^{th}$ element of A_p and $b_p(i,j)$ is the $(i,j)^{th}$ element of B_p.

The design objective is to find, without explicit knowledge of A_p and B_p, some control $u_p(t)$ such that the plant output vector $y_p(t)$ follows the output of the reference model

$$\begin{aligned} \dot{x}_m(t) &= A_m x_m(t) + B_m u_m(t) \\ y_m(t) &= C_m x_m(t) \end{aligned} \quad (5.24)$$

The model incorporates desired plant behavior and in many cases

$$dim[x_p(t)] \gg dim[x_m(t)] \quad (5.25)$$

5.6. CONTROL STRATEGY

The adaptive control law is given as [764]

$$u_p(t) = K_e(t)[y_m(t) - y_p(t)] + K_x(t)x_m(t) + K_u(t)u_m(t) \quad (5.26)$$

where $K_e(t)$, $K_x(t)$, and $K_u(t)$ are adaptive gains concatenated into a single matrix

$$K(t) = [K_e(t) \quad K_x(t) \quad K_u(t)] \quad (5.27)$$

Defining a vector $r(t)$ as

$$r(t) = \begin{bmatrix} y_m(t) - y_p(t) \\ x_m(t) \\ u_m(t) \end{bmatrix} \quad (5.28)$$

the control $u_p(t)$ can be written in compact form as

$$u_p(t) = K(t)r(t) \quad (5.29)$$

Thus, $u_p(t)$ is composed of the feedback term

$$K_e(t)e_y(t) \quad (5.30)$$

where $e_y(t)$ is the output error $[y_m(t) - y_p(t)]$, together with the feedforward component

$$K_x(t)x_m(t) + K_u(t)u_m(t) \quad (5.31)$$

The adaptive gains are obtained as a combination of the following integral and proportional gains [764]

$$\begin{aligned} K(t) &= K_p(t) + K_i(t) \\ K_p(t) &= [y_m(t) - y_p(t)]r^T(t)T_p, \quad T_p \geq 0 \\ \dot{K}_i(t) &= [y_m(t) - y_p(t)]r^T(t)T_i, \quad T_i > 0 \end{aligned} \quad (5.32)$$

It should be noted that for asymptotic tracking, the plant is required to be almost strictly positive real (ASPR), that is, there exists a positive definite constant gain matrix K_e, not needed for implementation, such that the closed-loop transfer function

$$G(s) = [I + G_p(s)K_e]^{-1}G_p(s) \quad (5.33)$$

is strictly positive real (SPR). It can be shown [764] that $m \times m$ MIMO system represented by a transfer function matrix $G(s)$ is ASPR if it

1. Is minimum phase, i.e. the numerator polynomial of $\mid G(s) \mid = \frac{z(s)}{p(s)}$ is stable.

2. Has relative degree of m or zero, i.e. the difference of the degrees of the denominator and the numerator polynomials of $\mid G(s) \mid = \frac{z(s)}{p(s)}$ is m or zero.

3. Has a positive definite high frequency gain, i.e. if the plant has the minimal realization $G(s) = C(sI - A)^{-1}B$, then $CB > 0$.

It can easily be seen that the ASPR conditions are in fact very restrictive, and most physical plants may not satisfy these conditions. However, as it will be shown, the linearized welding process satisfies the ASPR conditions by simple modification of the output equation.

5.6.2 Implementation of the DMRAC

Before implementing the adaptive control algorithm, we need first to determine whether the plant is ASPR or not. To this effect, consider again the linearized plant dynamics given by (5.10), which can be represented in the transfer function matrix form

$$G_p(s) = \frac{1}{den} \begin{bmatrix} g_{11} & g_{12} \\ g_{21} & g_{22} \end{bmatrix} \quad (5.34)$$

where

$$\begin{aligned} g_{11} &= L_s \pi r_w^2 s - (R_a + R_s + \rho \bar{x}_4) \pi r_w^2 \\ g_{12} &= -(C_1 + 2C_2 \rho \bar{x}_4 \bar{x}_5) \\ g_{21} &= (E_a - \rho \bar{x}_5) \pi r_w^2 \\ g_{22} &= \pi r_w^2 s + C_2 \rho \bar{x}_5^2 \end{aligned} \quad (5.35)$$

$$\begin{aligned} den = s\pi r_w^2 s^2 &+ (L_s C_2 \rho \bar{x}_5^2 + (R_a + R_s + \rho \bar{x}_4)\pi r_w^2)s \\ &+ (R_a + R_s - \rho \bar{x}_4)C_2 \rho \bar{x}_5^2 \\ &+ E_a C_1 + 2 E_a C_2 \rho \bar{x}_4 \bar{x}_5 - \rho C_1 \bar{x}_5 \end{aligned} \quad (5.36)$$

5.6. CONTROL STRATEGY

The determinant of $G(s)$ is

$$|G(s)| = \frac{\pi r_w^2}{den} \quad (5.37)$$

It is easy to see that the plant is minimum phase and its relative degree is two. However, the high frequency gain matrix is positive semi-definite resulting in non-ASPR configuration. To alleviate this problem and make the system ASPR, we use the negative of the arc voltage as the output. Then, the output equation becomes

$$\begin{bmatrix} -\delta y_1 \\ \delta y_2 \end{bmatrix} = \underbrace{\begin{bmatrix} E_a & -R_a \\ 0 & 1 \end{bmatrix}}_{C_p} \begin{bmatrix} \delta x_4 \\ \delta x_5 \end{bmatrix} \quad (5.38)$$

Since the linearized plant with the above modified output matrix is ASPR, the DMRAC can now be applied for the control of THE GMAW process. The closed-loop GMAW process with the DMRAC scheme is given in Figure 5.35. We first define the reference model inputs as the desired current I_d and the desired arc voltage V_{ad}. We chose A first-order reference model for both current and arc voltage and set its time constant to $\tau = 0.01$ sec. Thus, the reference model takes the form

$$G_m(s) = \frac{y_m(s)}{u_m(s)} = \frac{1}{\tau s + 1} \begin{bmatrix} 1 & 0 \\ 0 & 1 \end{bmatrix} \quad (5.39)$$

Since digital control is implemented, the continuous DMRAC was discretized as given in the following:

Reference Model Dynamics
The discretized form of the continuous reference model given in (5.39), using the sampling time of $T_s = 5000 Hz = 0.0002$ sec. is

$$\begin{aligned}
x_{m1}(kT_s + T_s) &= 0.9802 x_{m1}(kT_s) + 0.00019801 u_{m1}(kT_s) \\
y_{m1}(kT_s + T_s) &= 100 x_{m1}(kT_s) \quad (5.40) \\
x_{m2}(kT_s + T_s) &= 0.9802 x_{m2}(kT_s) + 0.00019801 u_{m2}(kT_s) \\
y_{m2}(kT_s + T_s) &= 100 x_{m2}(kT_s) \quad (5.41)
\end{aligned}$$

where u_{m1} and u_{m2} are the desired inputs to the reference model and y_{m1} and y_{m2} are the outputs of the reference model. In the actual implementation, we chose the desired $V_{arc} = u_{m1}$ and the desired $I = u_{m2}$.

Integral Adaptation Dynamics

Discretization of the adaptation laws was made using Backwards Rectangular Approximation.

$$K_p(kT_s) = e_y(kT_s)r^T(kT_s)T_p \qquad (5.42)$$
$$K_i(kT_s + T_s) = K_i(kT_s) + T_s e_y(kT_s)r^T(kT_s)T_i \qquad (5.43)$$
$$K_r(kT_s) = K_p(kT_s) + K_i(kT_s) \qquad (5.44)$$

where $e_y = [e_{y1} \ e_{y2}]^T$ is the error vector and the vector r^T is in the form of

$$r^T = [e_{y1} \ e_{y1} \ x_{m1} \ x_{m2} \ u_{m1} \ u_{m2}] \qquad (5.45)$$

and K_r is

$$K_r = \begin{bmatrix} K_{e11} & K_{e12} & K_{x11} & K_{x12} & K_{u11} & K_{u12} \\ K_{e21} & K_{e22} & K_{x21} & K_{x22} & K_{u21} & K_{u22} \end{bmatrix} \qquad (5.46)$$

Finally, the adaptive control takes the form of

$$S = u_{p1} = K_{e11}e_{y1} + K_{e12}e_{y2} + K_{x11}x_{m1} + K_{x12}x_{m2}$$
$$+ K_{u11}u_{m1} + K_{u12}u_{m2} \qquad (5.47)$$

$$V_{oc} = u_{p2} = K_{e21}e_{y1} + K_{e22}e_{y2} + K_{x21}x_{m1} + K_{x22}x_{m2}$$
$$+ K_{u21}u_{m1} + K_{u22}u_{m2} \qquad (5.48)$$

5.6.3 Experimental Results

The overall block diagram of the adaptive controller is shown in Figure 5.35. Welding parameters are as defined in Table 5.3. In addition, during the experiments, weld speed and wire-feed speed were set to 30 ipm and 350 ipm, respectively. In each experiment, the controller was turned on at 3 sec. During the experiments, the initial value of the open-circuit voltage V_{oc} was set to $28V$. Total weld time was 7 sec for each experiment. Detailed discussions about the experiments can be found in [765]. In the following figures, current and arc voltage responses and control signals (open-circuit voltage and wire-feed speed)

5.6. CONTROL STRATEGY

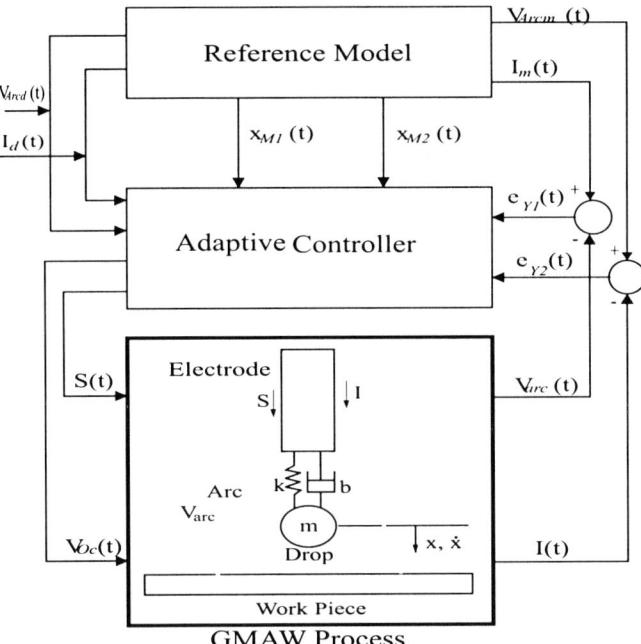

Figure 5.35: Closed-loop GMAW with DMRAC.

are shown for different current and arc voltage set points. Actual current in each experiment was computed by taking the mean of the actual current signal between 4 and 7 seconds. The weights for the adaptive gains were initially set to $T_i = T_p = diag(1e-5[1\,1\,1\,1\,1\,1])$ and finally tuned to $T_i = diag(1e-5[5\,5\,1\,100.11])$ and $T_p = diag(1e-5[1\,1\,1\,100.11])$.

Figures 5.36 and 5.37 show the actual current and arc voltage responses for the set points of 260A on current and 29V on arc voltage. Again, as seen from the figures both outputs reach closely to their desired values. However, the variations in the arc voltage are bigger in magnitude.

The actual current and arc voltage responses for the set points of 240A on current and 24V on arc voltage are shown in Figures 5.38 and 5.39. As the new set point moves further away from the initial steady-state value, the variations in the outputs increase. However, as seen from the figures both outputs reach closely to their desired values.

Figures 5.40 and 5.41 show the current and arc voltage responses when the desired values for current and arc voltage are set to 260A

Table 5.3: Welding Parameters Used During the Experiments

Weldment material	Flat stock mild steel
Weldment thickness	0.25"
Torch angle	0 degrees
Gas composition	85% Ar 15% CO2
Gas pressure	30.0 psi
Electrode diameter	0.045"
Electrode composition	AWS/ASME SFA5.18
Initial X-axis speed	30 ipm
Initial Y-axis speed	0.0 ipm
Power supply operating mode	Constant voltage
open-circuit voltage operating range	28V to 38V
wire-feed speed operating range	250 to 450 ipm
Contact tip to workpiece dist.	0.75"

and 29V, respectively for $3 < t \leq 5$ and 240A and 24V, respectively for $5 \leq t < 7$. As seen from the figures, the actual current and arc voltage reach closely to their desired values.

5.6. CONTROL STRATEGY

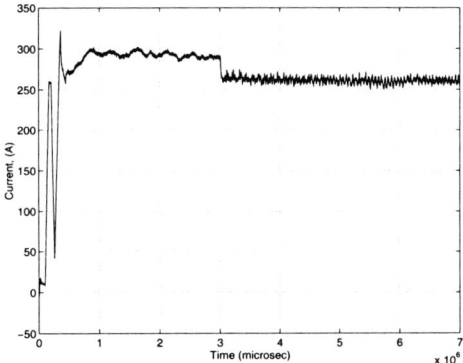

Figure 5.36: Current response, desired current = 260A, actual current = 260.3824A.

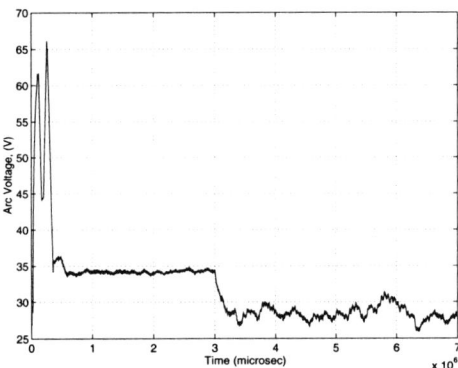

Figure 5.37: Arc voltage response, desired arc voltage = 29V, actual arc voltage = 28.6023V.

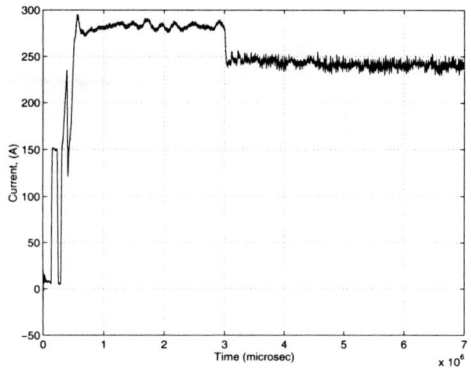

Figure 5.38: Current response, desired current = 240A, actual current = 240.9829A.

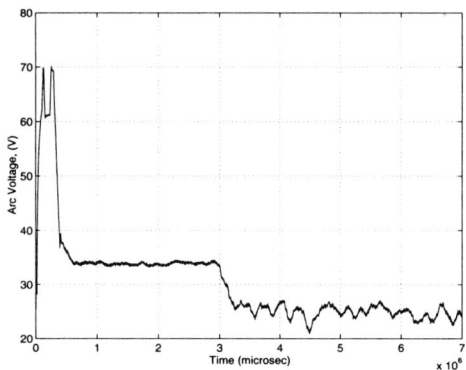

Figure 5.39: Arc voltage response, desired arc voltage = 24V, actual arc voltage = 24.8629V.

5.6. CONTROL STRATEGY

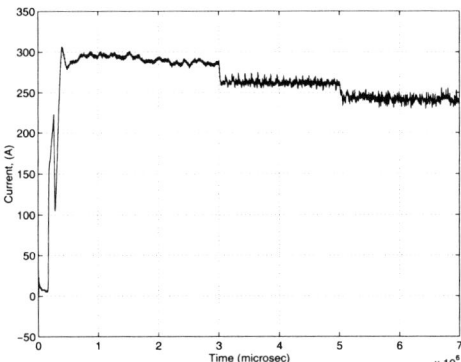

Figure 5.40: Current response, desired current = 260A, actual current = 261.9960A for $3 < t \leq 5$, desired current=240A, actual current=241.5589A for $5 < t \leq 7$.

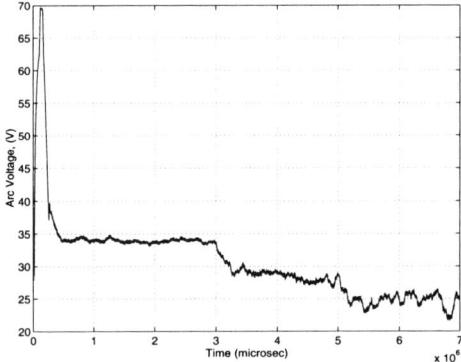

Figure 5.41: Arc voltage response, desired arc voltage = 29V, actual arc voltage = 28.0629V for $3 < t \leq 5$, desired arc voltage=24V, actual arc voltage=24.9090V for $5 < t \leq 7$.

5.7 Summary

In this case study, we have presented results on the control of the GMAW process, as part of a larger project aimed at developing methods for controlling GMAW processes, with an emphasis on the control of the mass and heat delivered by the process to the weld pool. Our focus was on regulation of current and arc voltage to prescribed set points. Experimental results showing the ability to achieve such regulation using classical controller design techniques were presented. First, a nonlinear model that depends on five empirical constants was described. Experimental data of steady-state current and arc voltage obtained over a range of operational conditions was presented. Analysis of the experimental data pointed out interesting insights regarding the selection of loop pairing variables for the purpose of feedback control, namely, that arc voltage is primarily affected by open-circuit voltage, while measured current is influenced by both open-circuit voltage and wire-feed speed. Next, the best value of each experimental parameter was determined relative to the error between the actual GMAW current and that predicted by the model, averaged over all the data. The best value of each parameter is also determined for each operating point. Contour plots of the model error show that the empirical "constants" are a function of the operating conditions of the system. Next, open-loop step response data was used to derive an empirical transfer matrix for the process. Results using classical control techniques to regulate the output of the GMAW were presented. A SISO controller for current was demonstrated. Then a multi-loop controller to simultaneously control current and measured arc voltage to desired set points was derived and demonstrated. The multi-loop controller is designed using the relative gain array, which suggests loop pairings that match wire-feed speed with current and open-circuit voltage with measured arc voltage. Experimental results, including a disturbance rejection test involving a step change in contact tube-to-workpiece distance, verify the effectiveness of the controller designs.

From these results, we can conclude that the control of the GMAW process through the use of a multi-loop PI control algorithm is possible. The multi-loop PI control algorithm also appears to be robust to disturbances in contact tip-to-workpiece distance. However, as noted, from contour plots of the model error, it is seen that the empirical "constants" of the fifth-order model are not constants but instead are

5.7. SUMMARY

a function of the operating conditions of the system. Thus, the empirical transfer function used for the multi-loop MIMO controller design will be dependent on the nominal operating conditions.

The final step of this case study was the design and implementation of a DMRAC for a GMAW process. Two models were used to control the system. First, a highly nonlinear fifth-order mathematical model for the GMAW process was used to describe the system dynamics. This model is the basis for the process-level control of current and arc voltage to desired set points. Through analysis of the model it was concluded that the current I and stick-out l_s are the dominant states, which determine the detachment properties. Therefore, the original fifth-order nonlinear process was reduced to a second order linearized plant. Second, a model of the heat and mass transfer from the process to the weld pool was presented. This model is used as the basis for prescribing the desired set points to be used by the process-level controller. Specifically, an optimization technique is used to find an input vector (contact tip-to-workpiece distance, open-circuit voltage, wire-feed speed and weld speed) and a corresponding output vector (physically possible current and stick-out) that result in maximum production rate for prescribed values of heat and mass transfer. Since the quality of welding is determined by the amount of mass G and heat H delivered to the workpiece, the relation between the mass and heat transferred to the work piece and the welding parameters was investigated. As a result, we defined the feasible region, where there exists a set of welding parameters $(CT, V_{oc}, S, R, I, l_s)$ that are physically possible for the given desired mass and heat values.

Then, the robust stability analysis was performed using the Kharitonov's theorem and it was found that the linearized system is stable for all the $G - H$ values considered, i.e., the system is stable in the feasible region. Then, the response of the system in the feasible region was investigated. It was found that the system is underdamped in most of the feasible region. For process-level control, the GMAW process was modeled as a two-input two-output second order system of differential equations. Current and arc voltage are the process outputs, which are controlled by open-circuit voltage and wire-feed speed, which are chosen as the process control inputs. Because of the nonlinear nature of the process, the system transfer function varies as a function of operating point. Thus, classical techniques as described above may not be adequate. For this reason, a DMRAC was designed and then implemented on the actual GMAW process. Experimental results show that the DMRAC is capable of controlling both the current and arc voltage. Hence, this new direct adaptive control algorithm allows the control of both current and arc voltage and consequently makes it possible to achieve high quality weldings in GMAW processes.

5.8 Classification of References by Section

Here, we provide a table containing the various references according to each section of this chapter. This will provide a ready reference to the interested reader to search for relevant references in each section.

Table 5.4: Section by Section List of References

Section	Reference Numbers
5.1 Introduction	[747]-[754]
5.2 Empirical Modeling of a GMAW Process	[748],[752]
5.3 SISO Current Control Using a PI Controller	
5.4 Multi-Loop Control of the GMAW Process	[754]-[756]
5.5 Adaptive Control of GMAW Process	[757]-[766]
5.6 Control Strategy	[764],[765]
5.7 Summary	[751],[757]

References List for Chapter 5

[747] K. Moore, M.A. Abdelrahman and D.S. Naidu. Gas metal arc welding control: Part II-control strategy. Nonlinear World Conference, 1998.

[748] K. Moore, D. Naidu, R. Yender and J. Tyler. Gas metal arc welding control: Part I-modeling and analysis. Nonlinear Analysis: Theory, Methods, and Applications. December, Pages 3101-3111, 1997.

[749] K. Moore and D.S. Naidu. Final report: Advanced welding project. Idaho State University, Pocatello, ID, 1997.

[750] K. Moore, J. Tyler, S. Ozcelik and D.S. Naidu. Classical control of gas-metal arc welding. American Control Conference. Philadelphia, PA, June, 1998.

[751] E.W. Reutzel, C. Einerson, J. Johnson, H Smartt, T Harmer and K. Moore. Derivation and calibration of a gas metal arc welding dynamic droplet model. The 4th International Conference on Trends in Welding Research. Gatlinburg, TN, June, 1995.

[752] R.F. Yender. Design, construction, and modeling of an automated gas metal arc welding facility for controller research. MS Thesis, Idaho State University, Pocatello, ID, 1997.

[753] J. Tyler. Model-based control of the gas metal arc welding process. MS Thesis, Idaho State University, Pocatello, ID, 1997.

[754] S. Ozcelik, K. Moore and D.S. Naidu. Mimo direct model reference adaptive control for gas metal arc welding. The 5th Interna-

tional Conference on Trends in Welding Research. Pine Mountain, GA. June, 1998.

[755] S. Ozcelik, K. Moore and D.S. Naidu. Application of MIMO direct adaptive control to gas metal arc welding. American Control Conference. June, 1998.

[756] D.E. Seborg, T.F. Edgar and D.A. Melichamp. Process, Dynamics, and Control. John Wiley and Sons. New York, NY. 1992.

[757] H.B. Smartt and C.J. Einerson. A model for heat and mass input control in gas metal arc welding. Welding Journal Pages 217-229, May, 1993.

[758] D.E. Henderson, P.V. Kokotovic, J.L. Schiano and D.S. Rhode. Adaptive control of an arc welding process. American Control Conference. Gatlinburg, TN. Pages 655-660, 1991.

[759] D.E. Henderson, P.V. Kokotovic, J.L. Schiano and D.S. Rhode. Adaptive control of an arc welding process. IEEE Control Systems Magazine. Pages 49-53, 1993.

[760] C.C. Doumanidis. Modeling and control of timeshared and scanned torch welding. Trans. of ASME, J. of Dynamics System, Measurement, and Control. Pages 387-394, 1994.

[761] G.C. Cook, K. Anderson and R.J. Barrett. Feedback and adaptive control in welding. 2nd Intl. Conf. on Trends in Welding Research. editor S. A. David and J. M. Vitek. Gatlinburg, TN. Pages 891-903, 1989.

[762] K.L. Moore, M.A. Abdelrahman and D.S. Naidu. Gas metal arc welding control: part II- control strategy. 2nd World Congress of Nonlinear Analysis. Athens, Greece. July, 1996.

[763] K.L. Moore, D.S. Naidu, M.A. Abdelrahman and A. Yesildirek. Advanced welding control project: Annual report FY96. Idaho State University. 1996.

[764] H. Kaufman, I. Bar-Kana and K. Sobel. Direct Adaptive Control Algorithms. Springer-Verlag. England, 1998.

REFERENCES LIST FOR CHAPTER 5

[765] K.L. Moore, D.S. Naidu, S. Ozcelik, B. Yender and J. Tyler. Advanced welding control project: Annual report FY97. Idaho State University, 1997.

[766] B.R. Barmish. New tools for robustness of linear systems. Macmillan Publishing Co., NY., 1994.

Chapter 6

Conclusions

Finally, some thoughts about the importance of control technology in automation of the welding process are presented.

6.1 Control Technology and Automation in Welding

In general, the welding process is such a complicated physical process (more so the Gas Metal Arc Welding (GMAW) process) that there is a real need for control strategies to be applied to welding in order to maintain desired weld quality. With this in mind, a National Materials Advisory Board Committee (NMABC) on Welding Controls, under contract with the Department of Defense (DoD) and the National Aeronautics and Astronautics Administration (NASA), was asked to "identify the variables in welding processes to ascertain where the existing control technology must be better understood and improved to ensure the necessary uniformity and reproducibility of structural welds" [767]. This is an excellent report that, although nearly 15 years old, is still valid and worth reading in order to understand the concerns of the welding community regarding the injection of control strategies into the welding processes. Some of the conclusions and recommendations of the Committee are worth noting:

1. There is a substantial opportunity for transforming welding technology in industry from an *experience-based* technology to a

scientific-based technology. Modeling of the welding process is still far from completion.

2. Research in welding should not be limited to traditional metallurgical and mechanical properties but should include welding controls with emphasis on interdisciplinary team work.

3. Development of on-line (real-time) sensor technology, utilizing the spectacular developments in solid-state technology, should be an integral part of modeling and control of the welding processes.

4. Welding process control research, both mission-oriented and exploration-oriented (basic) research, should be strongly encouraged by all agencies, government and private.

Also, there is a wealth of knowledge within the control community in the areas of modeling, analysis, and control. In particular, in the control area there are excellent results in nonlinear control, optimal control, adaptive control, and H_∞ optimal control, for both lumped-parameter (ordinary differential equations) and distributed parameter (partial differential equations) systems, that need to be explored for possible application to welding processes. For this exploration, there should be a strong interdisciplinary team effort towards modeling, sensing, and control of welding processes.

6.2 Main Issues and Outlook

During the last 15 years or so, major advances have been made in welding science in

1. understanding the physical process,

2. weld structure and properties,

3. sensor technology, and

4. control and automation.

However, there are still problems to be addressed. Modeling, which demands the complete understanding of the process, is not yet complete.

6.2. MAIN ISSUES AND OUTLOOK

There seems to be very good progress in sensor technology in both contact and non-contact types. There seems to be a real need for using the abundant wealth of knowledge available in the automatic control field for the welding process. In particular, advances made in control, in areas such as adaptive control, which normally does not require accurate modeling, in nonlinear control, which attacks the nonlinear model directly, and finally in intelligent control, using artificial neural networks (ANN), fuzzy logic, genetic algorithms, expert systems, etc., promise strong contributions. Also, see [768] for a good discussion on the current issues and problems in, and outlook for, welding science.

6.2.1 Welding in Space Research

The National Aeronautics and Space Administration (NASA), in their plans for space exploration for the next 30 years, identified "welding in space technology" as a key area of research, with goals to define welding in space requirements and capabilities and plan a technology program that will establish welding as a viable program for assembly, construction, and repair of structures in space [769].

6.2.2 Smart Robotic Welders and Manufacturing

The ultimate goal of arc welding seems to be to build *smart* robotic arc welding machines incorporating *intelligence*. Also, see [770] for a discussion on SmartWeld, a system for intelligent design and fabrication of welding components developed by Sandia National Laboratories, Albuquerque, NM.

The use of robots in welding industry is increasing. As argued in [771], in 1995 approximately 1600 arc welding robots were sold in the USA. Robotic welding machines offer many advantages such as repeatability, higher productivity, and better quality. Some of the recent works on robotic welding systems can be found in [772, 773, 774, 775, 776, 777, 778].

An interesting look at the welding system within the structure of a flexible manufacturing system can be found in [779]. Some of the innovative welding technologies are discussed in [780, 781, 782].

6.3 Classification of References by Section

Here, we provide a table containing the various references according to each section of this chapter. This will provide a ready reference to the interested reader to search for relevant references in each section.

Table 6.1: Section by Section List of References

Section	Reference Numbers
6.1 Control Technology and Automation in Welding	[767]-[771]
6.2 Main Issues and Outlook	[772]-[782]

References List for Chapter 6

[767] J. G. Bollinger. Control of Welding Processes. National Research Council. Washington, DC, 1987.

[768] S. A. David and T. DebRoy. Current issues and problems in welding science. Science, Volume 257, Pages 497-501. July, 1992.

[769] B. Newton-Montiel. Workshop launches welding in space research. Welding Journal, Pages 45-47. March, 1990.

[770] Anonymous Computer-based smart processes designed to increase U.S. manufacturer's quality. Welding Journal, Volume 75, Pages 18-19. October, 1996.

[771] D. D. Harwig. Weld parameter development for robot welding. Technical Paper - Society of Manufacturing, RP96-291. September, 1996.

[772] E. Craig. Increase robotic welding productivity by flying tip burn back. Robotics World, Volume 14(1), Pages 34-35. February, 1996.

[773] K. Dixon. Robotic Welding lies at the seat of productivity. Welding Review International, Volume 16(2), Pages 190-191. May, 1997.

[774] S. Adolfsson, A. Bahraim and C. Invgar. Sequential probability ratio test method for quality monitoring GMA welding. Proceedings of the International Modal Analysis Conference - IMAC., Pages 1622-1628. Sem Bethel, Orland, Florida. February, Pages 1622-1628, 1997.

[775] S. Adolfsson, A. Bahraim, G. Bolmsjo and I. Claesson. Automatic quality monitoring in robotised GMA welding using a repeated sequential probability ratio test method. International Journal for the Joining of Materials, Volume 9(1), Pages 2-8." March, 1997.

[776] A. stefan, B. Ali, B. Gunnar and C. Ingvar. Quality Monitoring in robotized spray GMA weld. International Journal for the Joining of Materials, Volume 10(1-2), Pages 3-23, August, 1998.

[777] D. Keith. Explaining robotic GMAW. Welding and Metal Fabrication, Volume 66(6), Pages 4pp. July, 1998.

[778] D. W. Kim, J. S. Choi and B. O. Nnaji. International Journal of Production Research, Volume 36(4), Pages 957-979. April, 1998.

[779] P. Drews and K. Fuchs. Welding Automation, Proc. of the 2nd Int. Conf. on Trends in Welding Research. S. A. David and J. M. Vitek, Gatlinburg, TN. May, Pages 105-107, 1989.

[780] D. Dipietro and J. Young. Pulsed GMAW helps John Deere meet fume requirements. Welding Journal. Vol 75, No 10, Pages 57-58. October, 1996.

[781] A. Blackman and V. Dorling. Technology advancements push pipeline welding productivity. Welding Journal (Miami, Fla). Volume 79(8), Pages 39-44, August, 2000.

[782] M. Pezzutti. Innovative welding technologies for the automotive Industry. Welding Journal. Volume 79(6), Pages 43-46, 2000.

Bibliography

Adam, G. and T. A. Siewert (1989). On-line arc welding data acquisition and analysis system. In: *Proceedings of the 2nd International Conference on Trends in Welding Research*. Gatlinburg, TN. pp. 979–983.

Adam, G. and T. A. Siewert (1990). Sensing of GMAW droplet transfer modes using an RE 100s-1 electrode. *Welding Journal* **69**(3), 103s–108s.

Adolfsson, S., A. Bahraim and C. Invgar (1997*a*). Sequential probability ratio test method for quality monitoring gma welding. In: *Proceedings of the International Modal Analysis Conference - IMAC*. (Sem Bethel, Ed.). Orland, Florida. pp. 1622–1628.

Adolfsson, S., A. Bahraim, G. Bolmsjo and I. Claesson (1997*b*). Automatic quality monitoring in robotised gma welding using a repeated sequential probability ratio test method. *International Journal for the Joining of Materials* **9**(1), 2–8.

Adolfsson, S., K. Ericsson and A. Greenberg (1996). Automatic detection of burn-through in gma welding using a parametric model. *Mechanical Systems & signal Processing* **10**(5), 633–651.

Aendenroomer, A. J. (1996). Weld Pool Oscillation for Penetration Sensing and Control. PhD thesis. Technical School of Delft University. Delft, The Netherlands.

Agapakis, J. E. and J. M. Katz (1989). An approach for knowledge-based sensor-controlled robotic welding. In: *Proceedings of the 2nd International Conference on Trends in Welding Research* (S. A. David and J. M. Vitek, Eds.). Gatlinburg, TN. pp. 935–939.

Agapakis, J. E., K. Masubuchi and N. Wittels (1984). General visual sensing techniques for automated welding fabrication. In: *Proceedings of the 4th International Conference on Robot Vision and Sensory Controls RoViSeC4*. London, UK. pp. 103–114.

Agapakis, J. E., N. Wittles and K. Masubuchi (1985). Automated visual weld inspection for robotic welding fabrication. In: *Proc. of the Intl. Conf. on Automation and Robotization in Welding and Allied Processes.* pp. 151–160.

Ainscough, D. M. (1987). Automatic Control of Weld Penetration. PhD thesis. The University of Liverpool. Liverpool, UK.

Alekseev, K. B., M. M. Fishkis and V. V. Fokin (1980). The adaptive control of a welding robot. *Welding Production(GB)* **27**(9), 4–8.

Alekseev, K. B., V. A. Afonin and M. M. Fishkis (1979). A two-channel control system for an adaptive welding robot. *Automatic Welding(GB)* **32**(12), 25–28.

Amin, M. and N.-Ahmed (1987a). Synergic control in MIG welding 2 - power-current controllers for steady state DC open arc operation. *Metal Construction* **19**(6), 331–340.

Amin, M. and N.-Ahmed (1987b). Synergic control in MIGwelding 1 - parametric relationships for steady DC open arc and short circuiting arc operation. *Metal Construction* **19**, 22–28.

Andersen, K. (1992). Studies and implementation of stationary models of the gas tungsten arc welding processes. Master's thesis. Vanderbilt University. Nashville, TN.

Andersen, K. (1993). Synchronous weld pool oscillation for monitoring and control. PhD thesis. Vanderbilt University. Nashville, TN.

Andersen, K. and R. J. Barnett an G. E. Cook (1988). Intelligent gas tungsten arc welding: Final report. Technical report. Vanderbilt University. Nashville, TN.

Andersen, K., G. E. Cook, L. Yizhang, D. S. Mathews and M. D. Randall (1989a). Modeling and control parameters for GMAW: short-circuiting transfer. In: *Advances in Manufacturing Systems:*

Integration and Processes (D. A. Dorfield, Ed.). Dearborn, MI. pp. 413–421.

Andersen, K., G. E. Cook, R. J. Barnett and A. M. Strauss (1997). Synchronous weld pool oscillation for monitoring and control. *IEEE Transactions on Industry Applications* **33**(2), 464–471.

Andersen, K., G. E. Cook, R. J. Barnett and E. H. Eassa (1989*b*). A class-H amplifier power source used as a high- performance welding research tool. In: *Proceedings of the 2nd International Conference on Trends in Welding Research* (S. A. David and J. M. Vitek, Eds.). Gatlinburg, TN. pp. 973–978.

Andersen, K., R. J. Barnett, J. F. Springfield and G. E. Cook (1992). Weldsmart: A vision-based system for quality control. Technical Report NASA Contract NAS8-37685. Vanderbilt University.

Anderson, R. C. (1993). *Inspection of Metals: Visual Examination*. American Society of Metals. Metals Park, OH.

Anonymous (1986). *Composition Controlled Sensing Technology*. U-Weld Automation. Rockville, MD.

Anonymous (1996). Computer-based smart processes designed to increase U.S. manufacturer's quality. *Welding Journal* **75**, 18–19.

Anonymous (1997*a*). Gas metal arc welding: transfer modes. *Welding Journal* **76**, 58–59.

Anonymous (1997*b*). How to calculate the cost of gas metal arc welding. *Welding Journal* **76**, 53–55.

Arata, Y. and K. Inoue (1972). Automatic control of arc welding by monitoring the molten pool. *Transactions of Japan Welding Research Institute (JWRI)* **1**(1), 99s–113s.

Arata, Y. and K. Inoue (1973). Automated control of arc welding (report II)-optical sensing of joint configuration. *Transaction. Jap. Welding Res. Institute (JWRI)* **2**(1), 87s–101s.

Arata, Y. and K. Inoue (1975). Automatic control arc welding (report IV). *Transactions Japan Welding Research Institute (JWRI)* **4**(2), 101s–104s.

Arata, Y., K. Inoue, K. Futamata and T. Toh (1979a). Investigation on welding arc sound-effect of welding method and welding condition of welding arc sound. *Transactions of Japan Welding Research Institute (JWRI)*.

Arata, Y., K. Inoue, M. Morita and G. Kawasaki (1976). Automatic control of arc welding (report V)-application of digital picture processing technique to automatic control. *Transaction of Japan Welding Research Institute (JWRI)* **5**(1), 77–85.

Arata, Y., K. Inoue, Y. Shibata, M. Tamaoki and H. Akashi (1979b). Automatic control of arc welding (report I)- algorithm for automatic selection of optimum welding condition. *Transactions of Japan Welding Research Institute (JWRI)* **8**(1), 1–11.

Araya, T. and S. Saikawa (1992). Recent activities on sensing and adaptive control of arc welding. In: *Proceedings of the 3rd International Conference on Trends in Welding Research* (S. A. David and J. M. Vitek, Eds.). Gatlinburg, TN. pp. 833–842.

Aström, K. J. and B. Wittenmark (1980). Self-tuning controllers based on pole-zero placement. *IEEE Proceedings*.

Aström, K. J. and B. Wittenmark (1995). *Adaptive Control*. Second ed.. Addison-Wesley Publishing Company, Inc.. Reading, MA.

Bachorski, A., M. J. Painter, A. J. Smailes and M. A. Wahab (1999). Finite-element prediction of distortion during gas metal arc welding using the shrinkage volume approach. *Journal of Materials Processing Technology* **92-93**, 405–409.

Baheti, R. S., K. B. Haefner and L. M. Sweet (1984). Operational performance of vision-based arc welding robot control systems. Technical report. General Electric Company. Schenectady, NY.

Baheti, R.S. (1985). Vision processing and control of robotic arc welding system. In: *Proceedings of the 24th IEEE Conference on Decision and Control*. Ft. Lauderdale, FL. pp. 1022–1024.

Banerjee, P., J. Liu and B. A. Chin (1992). Infrared thermography for non-destructive monitoring of weld penetration variations. In: *Proceedings of ASME Japan/USA Symposium Flex. Auto.*. San Francisco, CA. pp. 291–295.

Banerjee, P., S. Govardhan, H. C. Wikle and J. Y. Liu B. A. Chin (1995). Infrared sensing for on-line weld geometry monitoring and control. *ASME Transactions, Journal of Engineering for Industry* **117**, 323–330.

Barmish, B.R. (1994). *New tools for robustness of linear systems.* Macmillan Publishing Co.. NY, NY.

Barnett, R. J. (1993). Sensor development for multi-parameter control of gas tungsten arc welding. PhD thesis. Vanderbilt University. Nashville, TN.

Barnett, R. J., G. E. Cook, A. M. Strauss, K. Anderson and J. F. Springfield (1995a). A vision-based weld quality evaluation system. In: *Proceedings of the ASM 4th International Conference on Trends in Welding Research* (H. B. Smartt, J. A. Johnson and S. A. David, Eds.). Gatlinburg, TN. pp. 689–694.

Barnett, R. J., G. E. Cook, J. D. Brooks and A. M. Strauss (1995b). A weld penetration control system using synchronized current pulses. In: *Proceedings of the ASM 4th International Conference on Trends in Welding Research* (H. B. Smartt, J. A. Johnson and S. A. David, Eds.). Galtinburg, TN. pp. 727–732.

Barr, I. W. (1979). *Elementary Statistical Quality Control.* Dekker Publications. New York, NY.

Bates, B. E. and D. E. Hardt (1985). A real-time calibrated thermal model for closed-loop weld bead geometry control. *ASME Journal of Dynamic Systems, Measurement and Control* **107**, 25–33.

Beardsley, H. E., Y. M. Zhang and R. Kovacevic (1994). Infrared sensing of full penetration state in GTAW. *International Journal of Machine Tools and Manufacture* **34**(8), 1079–1090.

Beck, J.V. (1990). Inverse problems in heat transfer with applications to solidification and welding. In: *Proceedings of the 5th International Conference on Modeling of Casting and Welding Processes* (M. Rappaz, Ed.). Davos, Switzerland. pp. 503–515.

Begin, G. and J. P. Boillot (1983). Welding adaptive functions performed through infrared(ir) simplified vision schemes. In: *Proceedings of the 3rd International conference on Robot Vision and*

Sensory Controls RoViSeC3. Vol. 449. Cambridge, MA. pp. 328–337.

Bendant, J. S. and A. G. Piersol (1971). *Random Data: Analysis and Measurement Procedures.* Wiley-Interscience. New York, NY.

Bennett, A. P. and C. J. Smith (1976). Improving the consistency of weld penetration by feedback control. In: *Proceedings of Fabrication and Reliability Welded Process Plant.* The welding Institute. London, UK. pp. 13–19.

Bentley, A. E. and S. J. Marburger (1992). Arc welding penetration control using quantitative feedback theory. *Welding Journal* **71**(11), 397–404.

Best, R. (1993). *Phase-Locked Loops: Theory, Design and Applications.* Mc-Graw Hill. New York, NY.

Bingul, Z. and G. Cook (1999). Dynamic modeling of gmaw process. In: *Proceedings - IEEE International Conference on Robotics and Automation.* Piscataway, New Jersey. pp. 3059–3064.

Bjorgvinsson, J. B. (1992). Adaptive voltage control in gas tungsten arc welding. Master's thesis. Vanderbilt University. Nashville, TN.

Blackman, A. Stephen and V. David Dorling (2000). Technology advancements push pipeline welding productivity. *Welding Journal (Miami, Fla)* **79**(8), 39–44.

Blackmon, D. R. and F. W. Kearney (1983). A real-time quality approach to quality control in welding. *Welding Journal.*

Blodgett, O. W. (1987). Calculating cooling rates of arc spot welds. *Welding Journal* pp. 17–30.

Boillot, J. P., P. Cielo, G. Begin, C. Michel, M. Lessard, P. Fafard and D. Villemure (1985). Adaptive welding by fiber optic thermographic sensing: An analysis of thermal and instrumental considerations. *Welding Journal* **64**(7), 209s–217s.

Bollinger, J. G. (1987). Control of welding processes. Technical report. National Research Council. Washington, DC.

Bonser, G. and A. G. Parker (1999). Robotic gas metal arc welding of small diameter saddle type joints using multistripe structured light. *Optical Engineering* **38**(11), 1943–1949.

Bonvalet, J. C., Y. Launay and C. Philip (1985). Adaptive welding control using a CCD sensor. In: *Proc. of Intl. Conf. on Automation and Robotization in Welding and Allied Processes*. Intl. Inst. of Welding and Pergamon. Oxford, UK. pp. 365–367.

Boo, K. S. and H. S. Cho (1994). Determination of a temperature sensor location for monitoring weld pool size in GMAW. *Welding Journal* **73**(11), 265s–271s.

Boughton, P., G. Rider and C. J. Smith (1978). Feedback control of weld penetration. In: *Proceedings of 4th International Conference on Advances in Welding Processes*. Harrogate, England. pp. 203–215.

Boughton, P., G. Rider and G. J. Smith (1979). Towards the automation of arc welding. Technical report. CEGB Research, Central Electricity Generating Board. London, UK.

Brandi, S., C. Taniguchi and S. Liu (1991). Analysis of metal transfer in shielded metal arc welding. *Welding Journal* pp. 261s–270s.

Brogan, W. L. (1991). *Modern Control Theory*. Third ed.. Prentice Hall. Englewood Cliffs, NJ.

Brosilow, R. (1987). The new GMAW power supplies. *Welding Design & Fabrication* **60**(6), 22–28.

Brown, L. J. and S. P. Meyn (1995). Adaptive dead-time compensation. In: *Proceedings of the IEEE Conference on Decision and Control*. Vol. 4. New Orleans, LA. pp. 3435–3437.

Brown, L. J., S. P. Meyn and R. A. Weber (1995). Adaptive dead-time compensation with applications to a robotic arc welding system. Technical report. University of Illinois. Urbana, IL.

Brown, L., S. Meyn and R. Weber (1998). Adaptive dead-time compensationwith application to a robotic welding system. *IEEE Trasnactions on Control Systems Technology* **6**(3), 335–349.

Bruemmer, F. and R. Niepold (1987). Understanding and controlling arc welding processes by monitoring the melting pool with an optical sensor. In: *Proc. of the Intl. Workshop on Industrial Applications of Machine Vision and Machine Intelligence*. Tokyo, Japan. pp. 285–290.

Budai, P. and B.Torstensson (1985). A power source for advanced welding systems. In: *Proceedings of the EWI 1st International Conference on Advanced Welding System* (P. T. Houldcroft, Ed.). London, UK. pp. 424–434.

Burke, M. A., H. B. James and R. M. Wells (1987). The robotic adaptive welding system-RAWS. In: *Proceedings of First International Conference on Advanced Welding systems* (P.T. Houldcroft, Ed.). Welding Institute, Cambridge, UK. London, UK. pp. 31–40.

Cao, N. Z.n and P. Dong (1998). Modeling of gma weld pools wiht consideration of droplet impact. *Journal of Engineering Materials and Technology, Transactions of the ASME* **120**(4), 313–320.

Carlson, N. M. and J. A. Johnson (1986). Ultrasonic inspection of partially completed weld using pattern-recognition techniques. In: *Review of Progress in Quantitative Nondestructive Evaluation* (D. O. Thompson and D. E. Chimenti, Eds.). Plenum Publishing Co.. New York, NY. pp. 773–780.

Carlson, N. M. and J. A. Johnson (1988). Ultrasonic sensing of weld pool penetration. *Welding Journal* **67**(11), 293s–246s.

Carlson, N. M., J. A. Johnson and D. C. Kunerth (1990). Control of GMAW: Detection of discontinuities in the weld pool. *Welding Journal* **69**(7), 256s–263s.

Carlson, N. M., J. A. Johnson, E. D. Larsen, A. Van Clark, S. R. Schaps and C. M. Fortunko (1992). Ultrasonic sensing of GMAW Laser/EMAT defect detection system. In: *International Conference on ASM 3rd Trends in Welding Research* (S. A. David and J. M. Vitek, Eds.). Gatlinburg, TN. pp. 859–863.

Cary, H. B. (1989). *Modern Welding Technology*. Second ed.. Prentice Hall. Englewood Cliffs, NJ.

Castner, H. R. (1995). Gas metal arc welding fume generation using pulsed current. *Welding Journal* **74**(2), 59s–68s.

Castner, H. R. and D. Barborak (1991). Expert system for diagnosis of discontinuities in gas metal arc welds. Technical Report MR9101. Edison Welding Institute. Columbus, OH.

Chan, B., J. Pacey and M. Bibby (1999). Modelling gas metal arc weld geometry using artficial neural network technology. *Canadian Metallurgical Quarterly* **38**(1), 43–51.

Chandel, R. S. (1987). Mathematical modeling of melting rates for submerged arc welding. *Welding Journal* pp. 135s–140s.

Chao, Y. J. and Q. Xinhai (1999). Three-dimensional modeling of gas metal arc welding process. *Society of Manufacturing engineers* **MR99-164**, 1–6.

Chen, J.H. and L. Kang (1989). Investigation of the kinetic process of metal-oxygen reaction during shielded metal arc welding. *Welding Journal* pp. 245s–251s.

Chen, S. B., L.Wu, Q. L. Wang and Y. C. Liu (1997). Self - learning fuzzy neural networks and computer vision for control of pulsed GTAW. *Welding Journal* pp. 201s–209s.

Chen, W. H. and B. A. Chin (1990). Monitoring joint penetration using infrared sensing techniques. *Welding Journal* **69**(4), 181s–185s.

Chen, W. H., B. A. Chin and S. Nagarajan (1989). Infrared sensing for adaptive arc welding. *Welding Journal* p. 462s.

Chen, X. Q. and J. Lucas (1990). A fast system for control of narrow gap TIG welding. In: *Proc. of Intl. Conf. on Advances in Joining and Cutting Processes*. Harrogate, UK.

Chin, B. A., N. H. Madsen and J. S. Goodling (1983). Infrared thermography for sensing the arc welding process. *Welding Journal* **62**(9), 227s–234s.

Cho, H. S. (1992). Application of AI to welding process automation. In: *Proceedings of ASME Japan/USA Symposium Flex. Auto.*. pp. 303–308.

Choi, S. K., Y. S. Kim and C. D. Yoo (1999). Dimensional analysis of metal transfer in gma welding. *Journal of Physics D: Applied Physics* **32**(3), 326–334.

Choo, R. T., J. Szekely and R. C. Westoff (1990). Modeling of high-current arcs with emphasis on free surface phenomena in the weld pool. *Welding Journal* pp. 346s–361s.

Christensen, N., V.L. Davies and K. Gjermundsen (1965). Distribution of temperatures in arc welding. *British Welding Journal* **54**(2), 54–75.

Clark, D. E., C. Buhrmaster and H. B. Smartt (1989). Droplet transfer mechanisms in GMAW. In: *Proc. of 2nd Intl. Conference on Trends in Welding Research*. Gatlinburg, TN.

Clocksin, W. F., J. S. E. Bromley, P. G. Davey, A. R. Vidler and C. G. Morgan (1985). An implementation of model-based visual feed-back for robot arc welding of thin sheet steel. *International Journal of Robotics Research* **4**(1), 13–26.

Collard, J. F. (1988). Adaptive pulsed GMAW control: The digipulse system. *Welding Journal* **67**(11), 35–38.

Connor, L. P., Ed.) (1987). *Welding Handbook*. Vol. 1. Eigth ed.. American Welding Society. Miami, FL.

Conrardy, C. (1991). Control of GMAW with coaxial vision. PhD thesis. Ohio State University.

Cook, G. (1999). Robotic arc welding. *Society of Manufacturing Engineers* **AD99-109**, 1–20.

Cook, G. C., K. Anderson and R. J. Barrett (1989*a*). Feedback and adaptive control in welding. In: *Proceedings of the 2nd International Conference on Trends in Welding Research* (S. A. David and J. M. Vitek, Eds.). Gatlinburg, TN. pp. 891–903. Key Note Address.

Cook, G. C., K. Anderson and R. J. Barrett (1989*b*). Feedback and adaptive control in welding. In: *2nd Intl. Conf. on Trends in Welding Research* (S. A. David and J. M. Vitek, Eds.). Gatlinburg, TN. pp. 891–903.

Cook, G. E. (1968). Intrinsic thermocouple monitors welding. *Metals Progress* **93**, 176–180.

Cook, G. E. (1980a). Automated arc welding-return on investment. In: *Proceedings of Automated Arc Welding: How To Make The Right Decisions*. American Welding Society. Cleveland, OH. pp. 27–45.

Cook, G. E. (1980b). Feedback and adaptive control of process variable in arc welding. In: *Proceedings of an International Conference on Developments in Mechanised, Automated and Robotic Welding*. Welding Institute. Cambridge, MA. pp. P321–P329.

Cook, G. E. (1981a). Feedback and adaptive control in automated arc welding systems. *Metal Construction* **13**(9), 551–556.

Cook, G. E. (1981b). Microcomputer control of an adaptive positioning system for robotic arc welding. In: *Proceedings of IEEE IECI Conference on Applications of Mini and Microcomputers*. San Francisco, CA. pp. 324–329.

Cook, G. E. (1983a). Through-the-arc sensing for arc welding. In: *Proceedings of the 10th NSF Conference on Production Research and Technology*. pp. 141–151.

Cook, G. E. (1992). Computer-based control system for gas tungsten arc welding. In: *Japan/USA Symposium on Flexible Automation*. pp. 197–301.

Cook, G. E., A. M. Wells Jr, H. M. Floyd and R. L. McKeown (1982a). Analyzing arc welding signals with a micro-computer. In: *IEEE IAS Annual Meeting*. San Francisco, CA. pp. 1282–1288.

Cook, G. E. and P. C. Levick (1980). Microcomputer control of joint tracking system for arc welding. In: *Proceedings of the Winter Annual Meeting of ASME on Computer Applications in Manufacturing Systems* (W.R. De Vries, Ed.). Chicago, IL. p. 93=101.

cook, G. E., J. E. Maxwell, R. J. Barnett and A. M. Strauss (1997). Statistical process control application to weld process. *IEEE Transactions on Industry Applications* **33**(2), 454–463.

Cook, G. E., K. Andersen and R. J. Barnett (1990). Feedback and adaptive control in welding. In: *Recent Trends in Welding Science and Technology* (S. A. David and J. M. Vick, Eds.). Metals Park, OH. pp. 891–903.

Cook, G. E., K. Andersen and R. J. Barrett (1992). Computer-based control system for GTAW. In: *Proceedings of ASME Japan/USA Symposium Flex. Auto.*. pp. 297–301.

Cook, G. E., K. Andersen and R. J. Barrett (1993). Welding and bonding. In: *The Electrical Engineering Hand Book* (R. C. Dorf, Ed.). pp. 2223–2237. CRC Press. Boca Raton, FL.

Cook, G. E., K. Andersen, R. J. Barnett and J. F. Springfield (1991). Intelligent gas tungsten arc welding. In: *Automated Welding Systems in Manufacturing* (J. Weston, Ed.). Unknown. Cambridge, UK.

Cook, G. E., P.C. Levick, D. Welch and Jr. A.M. Wells (1982b). Distributed microcomputer control of an automated arc welding system. In: *IEEE IAS Conference*. pp. 1296–1302.

Cook, G. E., R. J. Barnett, A. M. Strauss and F. M. Thompson Jr. (1995). Statistical weld process monitoring with expert interpretation. In: *Proceedings of the ASM 4th International Conference on Trends in Welding Research* (H. B. Smartt, J. A. Johnson and S. A. David, Eds.). Gatlinburg, TN. pp. 739–744.

Cook, G.E. (1983b). Robotic arc welding: Research in sensory feedback control. *IEEE Transactions on Industrial Electronics* **IE-30**(3), 252–268.

Cook, P. L. (1985). Quality control systems for pipeline welding-a model and quantitative analysis. *Welding Journal* pp. 39–47.

Craig, E. (1988). The plasma arc process-a review. *Welding Journal* pp. 19–25.

Craig, E. (1996). Increase robotic welding productivity by flighting tip burn back. *Robotics World* **14**(1), 34–35.

Cram, L. E. (1984). A numerical model of droplet formation. In: *Proceedings of the 1983 International Conference on Computational Techniques and Applications* (J. Noyce and C. Fletcher, Eds.). Elsevier Science Publishers. Sydney, Australia.

Cullen, C. P. (1988). An adaptive robotic welding system using weldwire touch sensing. *Welding Journal* **67**(11), 17–21.

Cullison, A. (1999). Get that spatter under control. *Welding Journal* **78**(4), 43–45.

D. Barborak, C. Conrardy, B. Madignan and T. Paskell (1999). Through-arc process monitoring techniques for control of automated gas metal arc welding. In: *Proceedings - IEEE International Conference on Robotics and Automation.* Piscataway, New Jersey. pp. 3053–3058.

D. V. Nishar, J. L. Schiano, W. R. Perkins and R. A. Weber (1994). Adaptive control of temperature in arc welding. *IEEE Control Systems Magazine* **14**(4), 4–12.

Daggett, E. H. (1985). Feedback control to improve penetration of dip transfer MIG. In: *Proceedings of the EWI 1st International Conference on Advanced Welding Systems* (P. T. Houldcroft, Ed.). London, UK. pp. 282–285.

Datta, R., D. Mukherjee, K. L. Rohira and R. Veeraraghavan (1999). Weldability evaluation of high tensile plates using gmaw process. *Journals of Materials Engineering and performance* **8**(4), 455–462.

David, S. A. and T. DebRoy (1992). Current issues and problems in welding science. *Science* **257**, 497–501.

Davies, M. H., M. Wahab and M. J. Painter (2000). Investigation of the interaction of a molten droplet with a liquid weld pool surface: a computational and experimental approach. *Welding Journal (Miami, Fla)* **79**(1), 18s–23s.

D'Azzo, J. J. and C. H. Houpis (1995). *Linear Control System Analysis and Design: Conventional and Modern.* Fourth ed.. McGraw-Hill. New York, NY.

Deam, R. T. (1989). Weld pool frequency: A new way to define a weld procedure. In: *Proceedings of 2nd International Conference on Trends in Welding Research* (S. A. David and J. M. Vitek, Eds.). Gatlinburg, TN. pp. 967–971.

Deam, R. T. and P. N. Drew (1990). Relationship between arc light, current and arc length in TIG welding. In: *Proc. of Intl. Conf. on Advances in Joining and Cutting Processes*. Harrogate, UK.

Deam, R. T., S. W. Simpson and J. Haidar (2000). Semi-empirical model of the fume formation from gas metal arc welding. *Journal of Physics D: Applied Physics* **33**(11), 1393–1402.

DeLapp, G. E. Cook D. R., R. J. Barnett and A. M. Strauss (1995). Modeling and control parameters for GMAW, short- circuiting transfer. In: *Proceedings of the ASM 4th International Conference on Trends in Welding Research* (H. B. Smartt, J. A. Johnson and S. A. David, Eds.). Gatlinburg, TN. pp. 721–726.

Dereniak, E. L. and G. Crowe (1984). *Optical Radiation Detectors*. John Wiley & Sons. New York, NY.

Di, L., R. S. Chandel and T. Srikanthan (1999). Static modeling of gmaw process using artificial neural networks. *Materials and manufacturing processes* **14**(1), 13–35.

Dillenbeck, V. R. and L. Castagno (1987). The effects of various shielding gases and associated mixtures in GMA welding of mild steel. *Welding Journal* **66**, 45–49.

Dilthey, U. A. (1989). Sensor technique for welding robots: some of the developments and trends. Technical Report XI-1129-89. Intl. Inst. of Welding.

Dilthey, U. and J. Heidrich (1999). Using al-methods for parameter scheduling, quality control and weld geometry determination in gma-welding. *ISIJ International* **39**(10), 1067–1074.

Dilthey, U. and R. Killing (1987). Contribution to calculating the heat input in pulsed arc gas shielded metal arc welding. *Schweissen - Schneiden* **39**(10), E160–E162.

Dilthey, U., J. Heidrich and T. Reichel (1997). On-line quality control in gas metal arc welding using artificial neural networks. *Schweissen und Schneiden/ Welding & Cutting* **49**(2), E22–E24.

Dipietro, D. and J. Young (1996). Pulsed gmaw helps john deere meet fume requirements. *Welding Journal* **75**(10), 57–58.

Ditschun, A., B. Zajaczkowski and D. Dorling (1985a). Pulsed FM-GMA welding. In: *Proceedings of the Conference on Welding for Challenging Environments.* Toronto, Canada. pp. 21–29.

Ditschun, A., D. Dorling, A. G. Glover, B. A. Graville and B. Zajaczkowski (1985b). The development and application of pulsed FM-GMA welding. In: *Proceedings of EWI 1st International Conference on Advanced Welding system* (P. T. Houldcroft, Ed.). London, UK. pp. 321–329.

Dixon, B. F. (1989). Control of magnetic permeability and solidification cracking in welded nonmagnetic steel. *Welding Journal* pp. 171s–180s.

Dixon, K. (1997). Robotic welding lies at the seat of productivity. *Welding Review International* **16**(2), 190–191.

Doherty, J., S. J. Holder and R. Baker (1983). Computerised guidance and process control. In: *Proceedings of Third International Conference on Robot Vision and Sensory Controls RoViseC3* (D. P. Cassasent and E. L. Hall, Eds.). Vol. 449. Cambridge, MA. pp. 482–487.

Dorn, L., K. Moneni, H. Mecke and T. Rummel (1994). Process control in the arcing and short-circuit phases during metal-arc active gas build-up welding under carbon dioxide. *Welding and Cutting* **41**(6-7), 2–5.

Dornfeld, D. A. and M. Tomizuka (1984). Development of a comprehensive control strategy for gas metal arc welding. In: *Proceedings of 11th Conference Prod. Research and Technology.* Soc. of Manuf. Eng.. Pittsburgh,PA. pp. 271–275.

Dornfeld, D. A., M. Tomizuka and G. Langeri (1982). Modeling and adaptive control of arc welding processes. In: *Measurement, Control in Batch Manufacturing* (D. E. Hardt, Ed.). pp. 53–64. ASME. New York, NY.

Dorschu, K. E. (1968). Control of cooling rates in steel weld metal. *Welding Journal* **47**(2), 49s–62s.

Doumanidis, C. and Y.-M. Kwak (1999). Geometry modeling and adaptation by infrared and laser sensing in thermal manufacturing with material deposition. *American Society of Mechanical Engineers, Manufacturing Engineering Division* **10**, 573–580.

Doumanidis, C. C. (1988). Modeling and Control of Thermal Phenomena in Welding. PhD thesis. MIT. Cambridge, MA.

Doumanidis, C. C. (1992a). GMAW weld bead geometry: A lumped dynamic model. In: *Proceedings of the 3rd International Conference on Trends in Welding Research* (S. A. David and J. M. Vitek, Eds.). Gatlinburg, TN. pp. 63–67.

Doumanidis, C. C. (1992b). Hybrid modeling for control of weld dimensions. In: *Japan/USA Symposium on Flexible Automation.* pp. 317–323.

Doumanidis, C. C. (1994a). Modeling and control of timeshared and scanned torch welding. *Transactions of ASME, Journal of Dynamic System, Measurement and control* **116**, 387–394.

Doumanidis, C. C. (1994b). Multiplexed and distributed control of automated welding. *IEEE Control Systems* **14**(4), 13–24.

Doumanidis, C. C. and D. E. Hardt (1988). Multivariable adaptive control of thermal properties during welding. In: *Proceedings of the Winter Annual Meeting of the ASME on Control Methods for the Manufacturing Processes* (D. E. Hardt, Ed.). Chicago, IL. pp. 1–12.

Doumanidis, C. C. and D. E. Hardt (1989). A model for in-process control of thermal properties during welding. *Transactions of the ASME, Journal of Dynamic Systems, Measurement, and Control* **111**, 40–50.

Doumanidis, C. C. and D. E. Hardt (1990). Simultaneous in-process control of heat-affected zone and cooling rate during arc welding. *Welding Journal* **69**, 186s–196s.

Doumanidis, C. C. and D. E. Hardt (1991). Multivariable adaptive control of thermal properties during welding. *Transactions of the ASME, Journal of Dynamic Systems Measurement and control* **113**(1), 82–92.

Doumanidis, C.C. (1994c). Modeling and control of timeshared and scanned torch welding. *Trans. of ASME, J. of Dynamics System, Measurement, and Control* pp. 387–394.

Draugelates, U. and A. Schram (1999). Narrow-gap gas-shielded metal-arc welding of fine-grained structural steels in a vertical-up position. *Schweissen und Schneiden / Welding and Cutting* **51**(3), E42–E44.

Drews, P. and A. H. Kuhne (1985). An automatic welding system. In: *Proceedings of EWI First International Conference on Advanced Welding Systems* (P. T. Houldcroft, Ed.). Edison Welding Institute. London, UK. pp. 191–198.

Drews, P. and K. Fuchs (1989). Welding automation. In: *Proceedings of the 2nd International Conference on Trends in Welding Research* (S. A. David and J. M. Vitek, Eds.). Gatlinburg, TN. pp. 105–107.

Du, R., M. A. Elebestawi and S. M. Wu (1995a). Automated monitoring of manufacturing processes, part 1: monitoring methods. *Transactions of ASME, Journal of Engineering for Industry* **117**, 121–132.

Du, R., M. A. Elebestawi and S. M. Wu (1995b). Automated monitoring of manufacturing processes, part 2: applications. *Transactions of ASME, Journal of Engineering for Industry* **117**, 133–1141.

Dubovetskii, S.V. (1985). The application of statistical models of weld formation to the design and control of automatic arc welding conditions. In: *Proceedings of the EWI 1st International Conference on Advanced Welding System* (P. T. Houldcroft, Ed.). London, UK. pp. 210–236.

Dufour, M. and G. Begin (1983). Adaptive robotic welding using a rapid image pre-processor. In: *Proceedings of Third International Conference on Robot Vision and Sensory Controls RoViseC3* (D. P. Cassasent and E. L. Hall, Eds.). Vol. 449. Cambridge, MA. pp. 338–345.

Dufour, M. and P. Cielo (1984). Optical inspection for adaptive welding. *Application Optics* **23**(4), 271–275.

Dufour, M., X. Maldague and P. Cielo (1986). Environmental-noise analysis in active-vision systems for adaptive welding. *Proceedings of the SPIE, Optical Techniques for Industrial Inspection* **665**, 321–332.

DuPont, J. N. and A. R. Marder (1995). Thermal efficiency of arc welding processes. *Welding Journal* **74**(12), 406s–416s.

Dzelnitzki, D. (1999). Increasing the deposition volume or the welding speed - advantages of heavy duty mag welding. *Welding Research Abroad* **45**(3), 10–17.

Eagar, T., D. Hardt and J. Lang (1993). Decoupling of thermally based processes for multivariable control. Part of Joint INEL, MIT, ISU Proposal submitted to DOE Office of Basic Energy Sciences.

Eagar, T. W. (1992). Resistance welding: a fast, inexpensive and deceptively simple process. In: *Proceedings of the 3rd International Conference on Trends in Welding Research* (S. A. David and J. A. Vitek, Eds.). Gatlinburg, TN. pp. 347–351.

Eager, T. W. (1986). The physics and chemistry of welding processes. In: *in Advances in Welding Science and Technology* (S. A. David, Ed.). Metals Park, OH. pp. 291–298.

Edwards, W. R. (1985). Controlling weld metal volume improves productivity and profit. *Welding Journal* pp. 44–46.

Einerson, C. J., H. B. Smartt, J. A. Johnson, P. L. Taylor and K. L. Moore (1992). Development of an intelligent system for cooling rate and fill control in GMAW. In: *Proceedings of the ASM 3rd International Conference on Trend in Welding Research* (S. A. David and J. M. Vitek, Eds.). Gatlinburg, TN. pp. 853–857.

Essa, H. E., G. E. Cook and A. M. Wells (1983). A high performance welding source and its application. In: *IEEE-IAS-1983 Annual Meeting*. pp. 1241–1244.

Essers, W. G. (1976). Plasma with GMA welding. *Welding Journal* **55**, 394–400.

Essers, W. G. and M. R. M. Van Gompel (1984). Arc control with pulsed GMA welding. *Welding Journal* pp. 26–32.

Essers, W. G. and R. Walter (1981). Heat transfer and penetration mechanisms with GMA and plasma-GMA welding. *Welding Journal* **60**, 37s–42s.

et al, A. Matsunawa, Ed.) (1991). *Sensors and Control Systems for Arc Welding*. Vol. IIW Doc. XII-1220-91. Japan Welding Society. Tokyo, Japan.

et al., J. E. Agapakis (1986*a*). Joint tracking and adaptive robotic welding using vision sensing of the weld joint geometry. *Welding Journal*.

et al, R. B. Madigan (1986*b*). Computer based control of full penetration GTA welding using pool oscillation sensing. In: *Proceedings of a Conference on Computer Technology in Welding*. The Welding Institute, London, UK.

Faber, W. and D. Lindenau (1985). Taktile sensoren zur automatisierung in der schweib-technik. *ZIS-Mitteilungen* **27**(12), 1290–1296.

Fan, H. F. and R. Kovacevic (1998). Dynamic analysis of globular metal transfer in gas metal arc welding - a comparison of numerical and experimental results. *Journal Of Physics D: Applied Physics* **31**(20), 2929–2941.

Farson, D., C. Conrady, J. Talkington, K. Baker, Kerschbaumer and P. Edwards (1998). Arc initiation in gas metal arc welding. *Welding Journal* **77**(8), 315s–321s.

Farson, D. F., R. W. Richardson and R. J. Mayham (1986). Numerical simulation of feedback control of arc welding processes. In: *Proc. of the Intl. Conf. on Trends in Welding Research*. Gatlinburg, TN.

Feng, L., S. Chen, L. Liangyu and S. Li (1998). Static equilibrium model for the bead formation in high speed gas metal arc welding. *China Welding (English Edition)* **7**(1), 22–27.

Fenn, R. (1985). Ultrasonic monitoring and control during arc welding. *Welding Journal* pp. 18–22.

Fihey, J. L., P. Cielo and G. Begin (1983). On-line weld penetration measurement using an infrared sensor. In: *Proceedings of International Conference Welding in Energy-Related Projects.* Toronto, Canada. pp. 177–188.

Fourligkas, N. and C. Doumanidis (1994). Distributed parameter control of automated welding processes. In: *Proceedings of 2nd IEEE Mediterranean Symposium On New Direction in Control & Automation.* Maleme-Chania, Crete, Greece. pp. 113–119.

Francis, R. E., J. E. Jones and D. L. Olson (1990). Effect of shielding gas oxygen activity on weld metal microstructure of GMA welded microalloyed HSLA steel. *Welding Journal* pp. 408s–415s.

Frazz, U., R. Klier and P. Giese (1996). Process-integrated heat treatment as a quality assurance measure in the joining of cast iron and steel by welding. *Schweissen und Schneiden / Welding and Cutting* **48**(4), E71–E72.

Froehleke, N., H. Mundinger, S. Beineke, P. Wallmeier and H. Grotstollen (1997). Resonant transition swithcing welding power supply. In: *IECON Proceedings (Industrial Electronics Conference).* Los Alamitos, CA. pp. 615–620.

Fujimura, H. E., E. Ide and H. Inoue (1988). Estimation of contact tip-work piece distance in gas metal arc welding. *Welding International* **2**(6), 522–528.

Fujimura, H., E. Ide and H. Inoue (1987a). Joint tracking control sensor of GMAW: Development of method and equipment for position sensing in welding with electric arc signals (report 1). *Transactions of the Japan welding Society* **18**(1), 32–40.

Fujimura, H., E. Ide and H. Inoue (1987b). Weave amplitude control sensor of GMAW: Development of method and equipment for position sensing in welding with electric arc signals (report 2). *Transactions of the Japan Welding Society* **18**(1), 41–45.

Fujimura, H., E. Ide and H. Inoue (1994). Robot welding with arc sensing. In: *Sensors and Control Systems in Arc Welding* (H. Nomura, Ed.). Chap. 26, pp. 228–237. Chapman & Hall. London, UK. English Translation of the Original 1991 Japanese Edition.

Fujita, K. and T. Ishide (1994). Adaptive control of welding conditions using visual sensing. In: *Sensors and Control Systems in Arc Welding* (H. Nomura, Ed.). Chap. 18, pp. 160–167. Chapman & Hall. London, UK. English Translation of the Original 1991 Japanese Edition.

Galopin, M. and E. Boridy (1983). Une approche statistique du choix d'un mode operatoire de soudage. *Soudage et Techniques Connexes* **37**(11-12), 403–412.

Garland, J. G. (1974). Weld pool solidification control. *Metal Construction and British Welding Journal* **6**(4), 121–127.

Gellerman, M. J. (1984). How difficult is it to learn gas metal arc welding?. *Welding Journal* p. 41.

Gellie, R. W. (1983). Sensing for automated welding. In: *Proceedings of the 31st Annual Conference on Welding and Computers*. Sydney, Australia. pp. 193–200.

Ghosh, P. K., S. R. Gupta and H. S. Randhawa (1999). Characteristics and critically of bead on plate deposition in pulsed current vertical-up gmaw of steel. *International Journal for the Joining Materials* **11**(4), 99–110.

Gladkov, E. A. and I. A. Guslistov (1977). The dependence of the intensity of the radiant flux on the parameters of the weld pool in systems for the automatic control of penetration. *Automatic Welding* **30**, 5–8.

Glickstein, S. S. (1979). Arc modeling for welding analysis. In: *Proceedings of the International Conference on on Arc physics and Weld Pool Behavior*. London, UK. pp. 1–16.

Glickstein, S. S. (1982). Basic studies of the arc welding process. In: *Trend in Welding Research in The United States* (S. A. David, Ed.). pp. 3–49. American Society for Metals. Mata park, OH.

Glickstein, S. S. and E. Friedman (1984). Weld modeling applications. *Welding Journal* **63**, 38–42.

Goldak, J., V. Breiguine and N. Dai (1995). Computational weld mechanics: a progress report on ten grand challenges. In: *Proceedings* of the ASM 4th International Conference on Trends in Welding Research (H. B. Smartt, J. A. Johnson and S. A. David, Eds.). Gatlinburg, TN. pp. 5–11.

Goldberg, F. (1985). Inductive seam-tracking improves mechanized and robotic welding. In: *Proceedings of the Intl. Conf. on Automation and Robotization in Welding and Allied Processes*. Intl. Inst. of Welding and Pergamon. Oxford, UK.

Graham, G. M. and C. I. Ume (1995). Laser array generated ultrasound for weld quality control. In: *Proceedings of the ASM 4th International Conference on Trends in Welding Research* (H. B. Smartt, J. A. Johnson and S. A. David, Eds.). Gatlinburg, TN. pp. 677–681.

Greene, B. W. (1990). Arc current control of a robotic welding system: modeling and control system design. Technical Report DC-114-UILU-ENG-89-2227. Illinois Univ. at Urbana-Champaign, Coordinated science Lab.. Champaign, IL.

Grimble, M. J. (1994). *Robust Industrial Control: Optimal Design Approach for Polynomial Systems*. Prentice Hall. Englewood Cliffs, NJ.

Groom, K. N., S. Nagarajan and B. A. Chin (1990). Automatic single v groove welding utilizing infrared images for error detection and correction. *Welding Journal* pp. 441s–445s.

Gupta, M. M. and Sinha, N. K., Eds.) (1996). *Intelligent Control Systems: Theory and Applications*. IEEE Press. New York, NY.

Habib, S. (1987). Process control package designs, monitors, evaluates welding operation. *Welding Journal* pp. 69–70.

Haefner, K., B. Carey, B. Bernstein, K. Overton and M. D'Andrea (1988). Real time adaptive spot welding control. In: *Proceedings of the Winter Annual Meeting of the ASME on Control Methods for the Manufacturing Processes* (D. E. Hardt, Ed.). Chicago, IL. pp. 51–62.

Haidar, J. (1998). Analysis of the formation of metal droplets in arc welding. *Journal of Physics D: Applied Physics* **31**(10), 1233–1244.

Haidar, J. and J. J. Lowke (1996). Predictions of metal droplet formation in arc welding. *Journal of Physics D: Applied Physics* **29**(12), 2951–2960.

Hajossy, R., P. Pastava and I. Morva (1999). Ignition of a welding arc during a short-ciruit of melted electrodes. *Journal of Physics D: Applied Physics* **32**(9), 1058–1065.

Hale, M. B. and D. E. Hardt (1992). Multi-output process dynamics of GMAW: limits to control. In: *Proceedings of the ASM 3rd International Conference on Trends in Welding Research*. Gatlinburg, TN. pp. 1015–1020.

Halmoy, E. (1979). Wire melting rate, droplet temperature, and effective anode melting potential. In: *Proceedings of an International Conference on Arc Physics and Weld Pool Behavior*. The Welding Institute. London, UK. pp. 49–57.

Hang, M. and A. Okada (1993). Computation of GMAW welding heat transfer with boundary element method. *Advances in Engineering Software* **16**, 1–5.

Hardt, D. E. (1988). Measuring weld pool geometry from pool dynamics. In: *Proceedings of 3rd Conference on Modeling of Casting and Welding Process*. pp. 3–17.

Hardt, D. E. (1990). Modeling and control of welding processes. In: *Proceedings of the Fifth Conference on Modeling of Casting and Welding Processes* (M. Rappaz, M. Ozgu and K. W. Mahin, Eds.). The Minerals, Metals and Materials Society. Davos, Switzerland. pp. 287–303.

Hardt, D. E. (1992). Welding process modeling and re-design for control. In: *Proceedings of 4th US/Japan Symposium on Flexible Automation*. San Francisco, CA. pp. 275–281.

Hardt, D. E. (1993). Modeling and control of manufacturing processes: getting more involved. *Transactions of the ASME, Journal of Dynamic Systems, Measurement and control* **115**, 291–300.

Hardt, D. E. and J. M. Katz (1989). Ultrasonic measurements of weld pool penetration. *Welding Journal* **63**, 273s–281s.

Hardt, D. E., D. A. Garlow and J. B. Weinert (1985). A model of full penetration arc welding for control system design. *Transactions of the ASME, Journal of Dynamic systems, Measurement and Control* **107**, 40–46.

Hardt, D. E., D. A.Garlow and J. B. Weinert (1983). A model of full penetration arc welding for control system design. In: *Proceedings of the Winter Annual Meeting of the ASME on Control of Manufacturing Processes and Robotic Systems* (D. E. Heart and W. J. Book, Eds.). ASME, New York. pp. 121–135.

Harmer, T. M., K. L. Moore, H. B. Smartt, J. A. Johnson and E. R. Reutzel (1994). Modeling and simulation of a GMAW. Unpublished Work.

Hartman, D. A., D. R. Delapp, G. E. Cook and R. J. Barnett (1999). Intelligent control in arc welding. *Intelligent Engineering Systems Through Artificial Neural Networks* **9**, 715–725.

Harwig, D. D. (1996). Weld parameter development for robot welding. *Technical Paper - Society of Manufacturing*.

Hauptmann, P. (1991). *Sensors: Principles and Applications*. Prentice Hall. Englewood Cliffs, NJ.

Heald, P. R., R. B. Madigan, T. A. Siewert and S. Liu (1993). Mapping the droplet transfer modes for an ERIOOS-1 GMAW electrode. *Welding Journal* **73**, 38s– 44s.

Heiro, H. and T. H. North (1976). The influence of welding parameters on droplet temperature during pulsed arc welding. *Welding and Metal Fabrication* pp. 482–485,518.

Heller, W. A. (1983). Fundamentals and capabilities of automated welding systems. In: *Proceedings of the Conference on Automation and Robotics for Welding*. Indianapolis, IN. pp. 17–29.

Henderson, D. E. (1990). Adaptive control of an arc welding process. Technical Report Rep. DC-148, UILU-ENG-90-2220. University of Illinois. Urbana, IL.

Henderson, D. E., P. V. Kokotovic, J. L. Schiano and D. S. Rhode (1991a). Adaptive control of an arc welding process. In: *Proceedings of the American Control Conference*. Boston, MA. pp. 655–660.

Henderson, D. E., P. V. Kokotovic, J. L. Schiano and D .S. Rhode (1991b). Adaptive control of an arc welding process. In: *American Control Conference*. Gatlinburg, TN. pp. 655–660.

Henderson, D. E., P. V. Kokotovic, J. L. Schiano and D. S. Rhode (1993a). Adaptive control of an arc welding process. *IEEE Control Systems Magazine* **13**, 49–53.

Henderson, D.E., P.V. Kokotovic, J.L. Schiano and D.S. Rhode (1993b). Adaptive control of an arc welding process. *IEEE Control Systems Magazine* pp. 49–53.

Hermans, M. J. and G. Ouden Den (1999). Process behavior and stability in short circuit gas metal arc welding. *Welding Journal* **78**(4), 137s–141s.

Herold, H., G. Neubert, M. Zinke, U. Dilthey and A. Borner (1996). Research into gas metal-arc welding with pulsed arc on high-alloy steels. *Welding & Cutting* **48**(9), E182–E186.

Heuckroth, C. J. (1990). Wire feeders for robotic GMAW. *Welding Design & Fabrication* **63**(9), 47–49.

Hewitt, P. J. and A. A. Hirst (1993). A systems approach to the control of welding fumes at source. *Annals of Occupational Hygiene* **37**, 297–306.

Hirschfield, J. A. (1975). Welding aluminum, more on gas metal-arc welding. *Welding Journal* **54**, 28–30.

Holm, H. (1995). Manufacturing state modeling and its applications. In: *Proceedings of the ASM 4th International Conference on Trends in Welding Research* (H. B. Smartt, J. A. Johnson and S. A. David, Eds.). Gatlinburg, TN. pp. 665–675.

Holmes, J. G. and Resnick (1979). A flexible robot arc welding system. In: *Proceedings of the American Society of Mechanical Engineers*. pp. 1–15.

Hoyaux, M. H. (1968). *Arc Physics*. Springer-Verlag, New York.

Hughes, R. (1985). Arc guided robot plasma arc welding. In: *Proceedings of the 1st International Conference on Advanced Welding Systems*. The Welding Institute, London, UK.

Huissoon, J. P., D. L. Strauss, J. N. Rempel, S. Bedi and H. W. kerr (1994). Multi-variable control of robotic gas metal arc welding. *Journal of Materials Processing Technology* **43**(1), 1–12.

Huissoon, J. P., H. W. Kerr, S. Bedi, D. C. Wechman and W. P. Stefanuk (1992). Design of an integrated robotic welding system. In: *Proceedings of the 3rd International Conference on Trends in Welding Research* (S. A. David and J. M. Vitek, Eds.). Gatlinburg, TN. pp. 915–926.

Ide, E., M. Matsui, M. Matsumoto, T. Kawano, H. Iwabuchi and J. Wakiyama (1997). Development of arc simulation system on gas shielded metal arc welding process. *Technical Review - Mitubishi Heavy Industries* **34**(1), 1–4.

Inoue, K. (1080). Simple binary image processor and its application to automatic welding. In: *Proc. of Joint Automatic Control Conference*.

Inoue, K. (1980). Image processing for on-line detection of welding process (report 11) - binary processing for the image of arc welding process. *Transaction JWRI* **9**(1), 27–30.

inoue, K., H. Akashi, M. Tamaoki, Y. Shibata and H. Arata (1980). Automatic control of arc welding (report 11). *Transaction Japanese Welding Research Institute(JWRI)* **9**(1), 31–37.

Ishizaki, K. (1979). solidification of the molten metal pool and bead formation. In: *Proceedings of an International Conference on Arc Physics and Weld Pool Behavior*. London, UK. pp. 267–277.

Isidori, A. (1995). *Nonlinear Control Systems, Third Edition*. Springer-Verlag. Berlin.

Jackson, C. E. (1960a). The science of arc welding: Part i-definition of arc. *Welding Journal*.

Jackson, C. E. (1960b). The science of arc welding part ii-consumable-electrode welding arc. *Welding Journal.*

Jackson, C. E. (1960c). The science of arc welding part iii-what the arc does. *Welding Journal.*

Jaffery, W. S. (1992). Automated robotic variable polarity plasma arc welding (VPPAW) for the space station freedom project (SSFP). In: *Proceedings of the 3rd International Conference on Trends in Welding Research* (S. A. David and J. M. Vitek, Eds.). Gatlinburg, TN. pp. 943–947.

Jang, J.-S. R., C.-T Sun and E. Mizutani (1997). *Neuro-Fuzzy and Soft Computing.* Prentice Hall PTR. Upper Saddle River, NY.

Janots, M. (1978). Adaptive system of resistance welding control. In: *Fourth International Conference on Advances in Welding Processes.* Harrogate, England. pp. 239–247.

Jhaveri, P., W. G. Moffatt and C. M. Adams Jr. (1962). The effect of plate thickness and radiation on heat flow in welding and cutting. *Welding Journal* **41**(1), 12s–16s.

Johnson, C. A. and A. M. Sciaky (1966). System for controlling length of welding arc. US Patent 3236997.

Johnson, J. A. and N. M. Carlson (1988). Noncontact ultrasonic sensing of weld pools for automated welding. In: *Proceedings of 3rd International Symp. on Nondestructive Characterization of Materials.* Saarbrucken, FRG. pp. 854–861.

Johnson, J. A., H. B. Smartt, D. E. Clark, N. M. Carlson, A. D. Watkins and B. J. Lathcoe (1990). The dynamics of droplet formation and detachment in gas metal arc welding. In: *Proceedings of the Fifth International Conference on Modeling of Casting and Welding Processes.* Davos, Switzerland. pp. 139–146.

Johnson, J. A., M. Waddoups, H. B. Smartt and N. M. Carlson (1992). Dynamics of droplet detachment in GMAW. In: *Proceedings of the 3rd International Conference on Trends in Welding Research* (S. A. David and J. M. Vitek, Eds.). Gatlinburg, TN. pp. 987–997.

Johnson, J. A., N. M. Carlson and H. B. Smartt (1989). Detection of metal-transfer mode in GMAW. In: *Proceedings of ASM International conference on Recent Trends in Welding Science and Technology*. EG&G Idaho, Inc. Idaho Falls. pp. 377–381.

Johnson, J. A., N. M. Carlson and L. A. Lott (1988a). Ultrasonic wave propagation in temperature gradients. *Journal of Nondestructive Evaluation* **6**(3), 147–157.

Johnson, J. A., N. M. Carlson, H. B. Smartt and D. E. Clark (1991). Process control of GMAW: Sensing of metal transfer mode. *Welding Journal* **70**, 91s–99s.

Johnson, J. A., N. M. Carlson, R. T. Allemeier and D. G. Bannister (1988b). Ultrasonic and video computerized data acquisition for automated welding. In: *Proceedings of Computer technology in welding*. Cambridge, UK.

Jones, J. E. (1992). Weld parameter modeling. In: *Proceedings of the ASM 3rd International Conference on Trends in Welding Research* (S. A. David and J. M. Vitek, Eds.). Gatlinburg, TN. pp. 895–898.

Jones, L. A., T. W. Eagar and J. H. Lang (1988). Images of a steel electrode in ar-2% oxygen shielding during current gas metal arc welding.. *Welding Journal (Miami, Fla)* **77**(4), 135s–141s.

Jones, L. A., T. W. Eagar and J. H. Lang (1992). Investigations of drop detachment control in gas metal arc welding. In: *Proceedings of the 3rd International Conference on Trends in Welding Research* (S. A. David and J. M. Vitek, Eds.). Gatlinburg, TN. pp. 1009–1013.

Jones, L. A., T. W. Eagar and J. H. Lang (1998a). Dynamic model of drops detaching from a gas metal arc welding electrode. *Journal of Physics D: Applied Physics* **31**(1), 107–123.

Jones, L. A., T. W. Eagar and J. H. Lang (1998b). Magnetic forces acting on molten drops in gas metal arc welding. *Journal of Physics D: Applied Physics* **31**(1), 93–106.

Jones, S. B. and G. Starke (1985). Applications, advantages and approaches for image analysis in the control of arc welding. In: *Esprit*

'84: Status Report of Ongoing Work (J. Roukens and J. F. Renuart, Eds.). pp. 425–455. Elsveir Science Publishers. Amsterdam, The Netherlands.

Jones, S. B., J. Doherty and G. R. Salter (1977). An approach to procedure selection in arc welding. *Welding Journal* **56**(7), 19–31.

Jonsson, P. G., J. Szekely, R. B. Madigan and T. P. Quinn (1995). Power characteristics in GMAW: Experimental and numerical investigation. *Welding Journal* **74**(3), 93s–102s.

Jr, N. R. Corby (1985a). Machine vision algorithms for vision guided robotic welding. In: *Proceedings of 4th International Conference on Robot Vision and Sensory Controls, RoViseC5*. London, UK. pp. 137–147.

Jr, W. J. Kerth (1985b). Knowledge-based expert welding. In: *Proceedings of Robots9 Conference*. pp. 5–98–5–110.

Justice, J. F. (1983). Sensors for robotic arc welding. In: *Proceedings on AWS Conference on Automation and Robotics For Welding*. Indianapolis, IN. pp. 203–210.

Kang, J. and S. Rhee (2001). The statistical models for estimating the amount of spatter in the short circuit transfer mode of gmaw. *Welding Journal* pp. 1s–8s.

Kangsanant, T. and Z.H. Wang (1993). CAD-based expert system for robotic welding. In: *Proceedings of the 12th Triennial World Congress of the International federation of Automatic Control*. Sydney, Australia. pp. 303–306.

Kannatey-Asibu, Jr., E. (1987). Analysis of the GMAW process for microprocessor control of arc length. *Transactions of the ASME, Journal of Engineering for Industry* **109**, 172–176.

Karastojanov, D. N. and G. N. Nachev (1985). Adaptive control of industrial robots for arc welding. In: *Proc. of the 9th Triennial World Congress of the IFAC*. Budapest, Hungary. pp. 2405–2410.

Katz, J. M. and D. E. Hardt (1983). Ultrasonic measurement of weld penetration. In: *Proceedings of Control of Manufacturing*

Processes and Robot Systems (D. E. Hardt and W. J. Book, Eds.). ASME. New York. pp. 79–95.

Kaufman, H., I. Bar-Kana and K. Sobel (1998). *Direct Adaptive Control Algorithms*. Springer-Verlag. England.

Kawahara, M. (1983). Tracking control system for complex shape of welding groove using image sensor. In: *Proceedings of IFAC/IFIP Symp.-Real Time Digital Control Applications*. Guadalajara, Mexico. pp. 257–263.

Keith, D. (1998). Explaining robotic gmaw. *Welding and Metal Fabrication* **66**(6), 4pp.

Khan, M. A., N. H. Madsen, J. S. Goodling and B. A. Chin (1986). Infrared thermography as a control for the welding process. *Optical Engineering* **25**(6), 799–805.

Khosla, P. K., C. P. Neuman and M. Prinz (1985). An algorithm for seam tracking applications. *The International Journal of Robotics Research* **4**(1), 27–41.

Kim, D. W., J. S. Choi and B. O. Nnaji (1998). Robot arc welding opeartions planning with a rotating/tilting positioner. *International Journal of Production Research* **36**(4), 957–979.

Kim, I. S., A. Basu and E. Siores (1996*a*). Mathematical models for control of weld beam penetration in the gmaw process. *International Journal Of Advanced Manufacturing Technology* **12**(6), 393–401.

Kim, I. S., A. Basu and E. siores (1996*b*). Mathematical models for control of weld bean penetration in the GMAW process. *International Journal of Advanced Manufacturing Technology* **12**(6), 393–401.

Kim, I. S. and A. Basu (1998). Mathematical model of heat transfer and fluid flow in the gas metal arc welding process. *Journal of Materials Processing Technology* **77**(1-3), 17–24.

Kim, J. W. and S. J. Na (1991*a*). A study on an arc sensor for gas metal arc welding horizontal fillets. *Welding Journal* pp. 216s–221s.

Kim, J. W. and S. J. Na (1991*b*). Study on arc sensor algorithm for weld seam tracking in gas metal arc welding butt joints. *Proceedings of* the *Institution of Mechanical Engineers, Part B:Journal of ISSN* **205**(B4), 247–255.

Kim, J. W. and S. J. Na (1991c). A study on prediction of welding current in gas metal arc welding part 1: Modeling of welding current in response to change of tip-to-workpiece distance. *Proceedings of the Institution of Mechanical Engineers, Part B: Journal of Engineering Manufacture* **205**, 59–63.

Kim, J. W. and S. J. Na (1991d). A study on prediction of welding current in gas metal arc welding part 2: Experimental modeling of relationship between welding current and tip-to-workpiece distance and its application to weld seam tracking system. *Proceedings of the Institution of Mechanical Engineers Part B: Journal of Engineering Manufacture* **205**, 65–69.

Kim, J. W. and S. J. Na (1991e). study on the three dimensional analysis of heat and fluid flow in gas metal welding using boundary fitted coordinates. In: *Proceedings of Winter Annual meeting of the American society of Mechanical Engineers.* ASME. Atlanta, GA. pp. 159–173.

Kim, J. W. and S. J. Na (1995). A study on the effect of contact tube to workpiece distance on weld pool shape in gas metal arc welding. *Welding Journal* **74**(5), 141–152.

Kim, Y. S. and T. W. Eagar (1993). Analysis of metal transfer in gas metal arc welding. *Welding Journal* **72**(6), 269s–278s.

kim, Y. S., D. M. McEligot and T. W. Eagar (1991). Analysis of electrode heat transfer in gas metal arc welding. *Welding Journal* p. 20s.

Kjeld, F. (1991). Gas metal arc welding for the collision repair industry. *Welding Journal* p. 39.

Kohn, M. L., S. A. Ramsay, D. T. Damon and B. W. Folkening (1985). A robotic cell for welding large aluminum structures using a two-pass optical system. In: *First International Conference on Advanced Welding systems* (P.T. Houldcroft, Ed.). Welding Inst, Abingdon, Cambridge, UK. pp. 17–29.

Kondoh, K. and T. Ohji (1998). Algorithm based on convex programming method for optimum heat input control in arc welding. *Materials Transactions* **39**(3), 420–426.

Koseeyaporn, P., G. Cook and M. A. Strauss (2000). Adaptive voltage control in fusion arc welding. *IEEE Transactions on Industry Applications* **36**(5), 1300–1307.

Kotecki, D. J., D. L. Cheevr and D. G. Howden (1972). Mechanism of ripple formation during weld solidification. *Welding Journal* **51**, 386s–391s.

Kou, S. and Y. Le (1985). Improving weld quality by low frequency arc oscillation. *Welding Journal* pp. 51–55.

Kovacevic, R. and Y. M. Zhang (1994a). Adaptive control of GTA welding of sheet metals. In: *Proceedings of the Fifth International Welding Computerization Conference*. Golden, CO. pp. 38–48.

Kovacevic, R. and Y. M. Zhang (1994b). Model-based adaptive vision control of weld pool area. In: *Proceedings of the American Control Conference*. Baltimore, MD. pp. 313–317.

Kovacevic, R. and Y. M. Zhang (1995a). Machine vision recognition of weld pool in gas tungsten arc welding. *Proceedings of Institution of Mechanical Engineers, Part B: Journal of Engineering Manufacture* **209**(B2), 141–152.

kovacevic, R. and Y. M. Zhang (1995b). Weld pool sensing and control: 2D shape 3D surface. In: *Proceedings of the International Symposium in Materials Science and Technology*. Harbin, China. pp. 379–384.

Kovacevic, R. and Y. M. Zhang (1996a). On-line measurement of weld fusion zone state using weld pool image and neurofuzzy model. In: *Proceedings of the 19 IEEE International Symposium on Intelligent Control*. Dearborn, MI. pp. 307–312.

Kovacevic, R. and Y. M. Zhang (1996b). Sensing free surface of arc weld pool using specular reflection: principle and analysis. *Proceedings of Institution of Mechanical Engineers, PartB Journal of Engineering Manufacturing* **210**(6), 553–564.

Kovacevic, R. and Y. M. Zhang (1997a). Real-time image processing for monitoring of free weld pool surface. *ASME Transactions, Journal of Manufacturing Science and Engineering for Industry.*

Kovacevic, R. and Yu M. Zhang (1997b). Neurofuzzy model-based weld fusion state estimation. *IEEE Control Systems* pp. 30–42.

Kovacevic, R., Y. M. Zhang and H. E. Beardsley (1993a). Weld penetration control with infrared sensing and neural network technology. In: *Proceedings of the International Conference on Modeling and Control of Joining Processes.* Orlando, FL. pp. 393–400.

Kovacevic, R., Y. M. Zhang and S. Ruan (1993b). Three-dimensional measurement of weld pool surface. In: *Proceedings of the International Conference on Model and Control of Joining Processes.* Orlando FL. pp. 600–607.

Kovacevic, R., Y. M. Zhang and S. Ruan (1995). Sensing and control of weld pool geometry for automated GTA welding. *ASME Transactions, Journal of Engineering for Industry* **117**(2), 210–222.

Kovacevic, R., Y. M. Zhang, E. Liguo and H. Beardsley (1996a). Dynamic analysis metal transfer process for GMAW control. *ASM Journal of Engineering Materials and Technology.*

Kovacevic, R., Y. M. Zhang, E. Liguo and H. E. Beardsley (1996b). Dynamics of droplet geometry during metal transfer in gmaw - a model for process control. In: *ASME International Mechanical Engineering Congress and Exposition.* Atlanta, Georgia. pp. 143–144.

Kovacevic, R., Y. M. Zhang, L. Li and H. Beardsley (1996c). Sensing and control weld geometrical appearance. In: *Proceedings of the 26th Conference on Production Engineering.* Budva, Yugoslavia.

Kovacevic, R., Z. N. Cao and Y. M. Zhang (1996d). Role of welding parameters in determining the geometrical appearance of weld pool. *ASME Transactions, Journal of Engineering Materials and Technology* **118**, 589–596.

Kuhne, A. H., H.B. Cary and F. B. Prinz (1987). An expert system for robotic arc welding. *Welding Journal* p. 21.

Kuk, K. A. (1985). Determining acceptable joint mislocation in systems without adaptive control. *Welding Journal* pp. 65–66.

Kumar, S. and S. C. Bhaduri (1994). Three dimensional finite element modeling of gas metal arc welding. *Metallurgical and Materials Transactions B: Process Metallurgy and Materials* **25**(3), 435–441.

Kumar, S. and S. C. Bhaduri (1995). Theoretical investigation of penetration characteristics in gas metal-arc welding finite element method. *Metallurgical and materials Transactions B: Process Metallurgy and Material* **26B**(3), 611–624.

Kuvin, B. F. (1989). Guided robots weld army tanks. *Welding Design and Fabrication* **62**(9), 41–44.

Kwak, Y. M. and C. Doumanidis (1999). Solid freeform fabrication by gma welding: Geometry modeling, adaption and control. *American Society of Mechanical Engineers, Manufacturing Engineering Division* **10**, 49–56.

L. Di, S. Yonglun and Y. Feng (2000). On line monitoring of weld defects for short-circuit gas metal arc welding based on the self-organize feature map neural networks. In: *Proceedings of the International Joint Conference Neural Networks*. Piscataway, New Jersey. pp. 239–244.

Lamkalapalli, K. N., J. F. Tu, K. H. Leong and M. Gartner (1999). Laser weld penetration estimation using temperature measurements. *Journal Of Manufacturing Science and Engineering* **121**(2), 179–188.

Lancaster, J. F., Ed.) (1986). *The Physics of Welding*. Second ed.. Pergamon Press. Oxford, UK.

Langari, G. and M. Tomizuka (1988). Fuzzy linguistic control of arc welding. In: *Proceedings of ASME Winter Annual Meeting on Sensors and Controls for Manufacturing*. Chicago, IL. pp. 157–162.

Lanzetta, M., M. Santochi and G. Tantussi (2001). On-line control of robotized gas metal arc welding. *CIRP Annals - Manufacturing Technology* **50**(1), 13–16.

Lee, C. W. and S. J. Na (1996a). Study on the influence of reflected arc light on vision sensors for welding automation. *Welding Journal* **75**(12), 379s–387s.

Lee, C. W. and S. J. Na (1996b). Study on the influence of reflected arc light on vision sensors for welding automation. *Welding Journal (Miami, Florida)* **75**(12), 379s–387s.

Lee, J. I. and S. Rhee (2000). Prediction of process parameters for gas metal arc welding by multiple regression analysis. In: *Proceedings of the Institution of Mechanical Engineers, Part B: Journal of Engineering Manufacture*. London, England. pp. 443–449.

Lesnewich, A. (1958). Control of melting rate and metal transfer in gas-shielded metal-arc welding part 1: Control of electrode melting rate. *Welding Journal* **37**(8), 343s–353s.

Lesnewich, A. (1991). Technical commentary: Observations regarding electrical current flow in the gas metal arc. *Welding Journal* pp. 171s–172s.

Lewis, F. L. and V.L. Syrmos (1995). *Optimal Control, Second Edition*. John Wiley & Sons. New York, NY.

Li, P. J. and Y. M. Zhang (2000). Analysis of an arc light mechanism and its application in sensing of the gtaw process. *Welding Journal (Miami, Fla)* **79**(9), 252s–260s.

Li, Y., L. Wu, D. Cheng and J. E. Middle (1992). Machine vision analysis of welding region and its application to seam tracking in GTAM and GMAW. In: *Proc. of the 3rd Intl. Conf. on Trends in Welding Research* (S. A. David and J. M. Vitek, Eds.). Gatlinburg, TN. pp. 1021–1025.

Lim, T. G. and H. S. Cho (1993). Estimation of weld pool sizes in GMA welding process using neural networks. *Proceedings of Institution of Mechanical Engineers Part 1, Journal of Systems and Control Engineering* **207**(1), 15–26.

Limmaneevichitr, C. and S. Kou (2000). Experiments to simulate effect of marangoni convection on weld pool shape. *Welding Journal (Miami, Fla)* **79**(8), 231s–237s.

Lin, M. and T. W. Eagar (1985). Influence of arc pressure on weld pool geometry. *Welding Journal* **64**, 163s–169s.

Lin, T. T., K. Groom, N. H. Madsen and B. A. Chin (1986). Infrared sensing techniques for adaptive robotic welding. In: *Proceedings of the Conference on Modeling and Control of Casting and welding Processes*. Santa Barbara, CA. pp. 19–31.

Linden, G. and G. Lindskog (1980). A control system using optical sensing for metal-inert gas arc welding. In: *Proceedings of TWI Conference*.

Linden, G., G. Lindskog and L. Nilsson (1980). A control system using optical sensing for metal-inert gas arc welding. In: *Proceedings of an International Conference on Developments in Mechanized, Automated and Robotic Welding*. London, UK. pp. P17–1–P17–6.

Liu, S. and T. A. Siewert (1989). Metal transfer in gas metal arc welding: droplet rate. *Welding Journal* **68**(2), 52s–58s.

Liu, S., T. A. Siewert, G. A. Adam and H.G Lan (1990). Arc welding process control from current and voltage signals. In: *Proceedings of The Welding Institute Conference on Computer Technology in Welding*. pp. 26–35.

Ljung, L. (1987). *System Identification: Theory for the User*. Prentice Hall. Englewood Cliffs, NJ.

Lott, L. A., J. A. Johnson and H. B. Smartt (1983). Real-time ultrasonic sensing of arc welding processes. In: *Proc. of 1983 Symp. on Nondestructive Evaluation Applications and Materials*. Metals Park, OH. p. 13=22.

L.Schiano, J., D. E. Henderson, J. H. Ross and R. A.Weber (1991a). Image analysis of puddle geometry and cooling rate for gas metal arc welding control. In: *Proc. of the Inst. of Soc. Opt. Eng., Midwest Tech. Conference*. Chicago, IL.

L.Schiano, J., J. H. Ross and R. A.Weber (1991b). Modeling and control of puddle geometry in gas metal arc welding. In: *Proceedings of the 1991 American Control Conference*. Boston,MA. pp. 1044–1049.

Lu, M. J. and S. Kou (1989*a*). Power inputs in gas metal arc welding of aluminum-part 1. *Welding Journal* pp. 382s–388s.

Lu, M. J. and S. Kou (1989*b*). Power inputs in gas metal arc welding of aluminum-part 2. *Welding Journal* pp. 452s–456s.

Lucas, W. and R. S. Mallet (1975). Automatic control of penetration in pulsed TIG welding. Technical report. The Welding Institute. London, UK.

Luijendijk, T., J. D. Zeeuw and M. P. Spikes (1996). Calculation of the electrical resistance between contact tube and welding wire during gma welding based on measurement of the contact force. *International Journal for the Joining of Materials* **8**(1), 1–4.

Lukens, W. E. and R. A. Morris (1982). Infrared temperature sensing of cooling rates for arc welding control. *Welding Journal* **61**(1), 27–33.

Madigan, R. B. (1987). Ways to keep torches in seams. *Welding Design and Fabrication* **60**(10), 48–50.

Madigan, R. B. (1994). Control of gas metal arc welding using arc light sensing. PhD thesis. Colorado School of Mines. Golden, CO.

Madigan, R. B., T. P. Quinn and T. A. Siewert (1992). Sensing droplet detachment and electrode extension for gas metal arc welding. In: *Proceedings of the 3rd International Conference on Trends in Welding Research* (S. A. David and J. M. Vitek, Eds.). Gatlinburg, TN. pp. 999–1008.

Madsen, N. H. and B. A. Chin (1984). Automatic welding: Infrared sensors for process control. In: *Proceedings of 11th Conference on Production Research and Technology*. PA, Society of Manufacturing Engineers. Pittsburgh. pp. 277–285.

Manley, T. D. and T. E. Doyle (1992). Arc hydrogen monitoring for synergic GMAW. In: *Proceedings of the ASM 3rd International Conference on Trends in Welding Research* (S. A. David and J. M. Vitek, Eds.). Gatlinburg, TN. pp. 1027–1030.

Manz, A. F. (1990). The dawn of gas metal arc welding. *Welding Journal* pp. 67–68.

Maqueira, B., C. Umeagukwu and J. Jarzynski (1987). Robotic seam tracking of weld joints through the use of ultrasonic sensors. In: *Proceedings of Intl. Workshop on Industrial Applications of Machine Vision and Machine Intelligence*. Tokyo, Japan. pp. 291–296.

Martin, T. E. (1995). *Process Control: Designing Processes and Control Systems for Dynamic Performance*. McGraw Hill. New York, NY.

Masmoudi, R. (1992). Process decoupling for simultaneous control of heat affected zone and width in GMAW. PhD thesis. Department of Mechanical Engineering, MIT.

Masmoudi, R. and D. E. Hardt (1992a). High-frequency torch weaving for enhanced welding controllability. In: *Proceedings of the ASM 3rd International Conference on Trends in Welding Research* (S. A. David and J. M. Vitek, Eds.). Gatlinburg, TN. pp. 931–936.

Masmoudi, R. and D. E. Hardt (1992b). Multivariable control of geometric and thermal properties in GTAW. In: *Proceedings of 3rd ASM Conference Trends welding Res.*. Gatlinburg,TN.

Masubuchi, K., D. E. Hardt, H. M. Paynter and W. Unkel (1981). Improvement of reliability of welding by in-process sensing and control. In: *Proceedings of Trends in Welding Research in The United State*. New Orleans, LA. pp. 667–688.

Mathews, D. S. (1988). GMAW short circuiting transfer: Stabilities, instabilities, and joule heating model. Master's thesis. Vanderbilt University. Nashville, TN.

Matsunawa, A. (1992). Modeling of heat and fluid flow in arc welding. In: *Proceedings of the ASM 3rd International Conference on Trends in Welding Research* (S.A. David and J. M. Vitek, Eds.). Gatlinburg, TN. pp. 3–16.

Matteson, M. A., R. A. Morris and D. Raines (1992). An optimal artificial neural network for GMAW arc acoustic classification. In: *Proceedings of ASM 3rd International Conference on Trends in Welding Research* (S. A. David and J. M. Vitek, Eds.). Gatlinburg, TN. pp. 1031–1035.

Matthes, Klaus-Juergen, Mario Kusch, Helmut Roth. Steffen Mueller and Franlk Wuest (1999). Control response of pulsed power sources for gas-shielded metal-arc welding when the contact-tube distance is changed. *Schweissen und Schneiden / Welding and Cutting* **51**(9), 201–204.

Maual, G. P., R. Richardson and B. Jones (1996). Statistical process control applied to gas metal arc welding. *Computers and Industrial Engineering* **31**(1-2), 253–256.

Maul, P. G., R. Richardson and B. Jones (1996). Statistical process control applied to gas metal arc welding. *Computers And Industrial Engineering* **31**(1-2), 253–256.

Maxwell, J. E. (1990). Statistical process control for arc welding processes. Master's thesis. Vanderbilt University. Nashville, TN.

McCampbell, W. M., G. E. Cook, L. E. Nordholt and G. J. Merrick (1966). The development of a weld intelligence system. *Welding Journal* **45**(3), 139s–144s.

McGlone, J. C., J. Doherty and S. B. Jones (1979). Developments in arc welding procedure selection. In: *Proceedings of ANS Conference on Welding and Fabrication in the Nuclear Industry*. Nuclear Energy Society. London, UK. pp. 283–289.

Mecke, H., W. Fischer, I. Merfert, S. Nowak, U. Dilthey, U. Reisgen, L. Stein and H. Bachem (1998). Combined primary and secondary-clocked, computer-controlled welding power supplies with higher dynamice ratio. *Schweissen and Schneiden Welding & Cutting*. **50**(4), E67–E69.

Metko, G. N. (1984). Pulsed spray transfer welding. In: *Proceedings of Conference on Sheet Metal Welding Conference The Latest Technology for Sheet Metal Joining*. American Welding Society. Detroit, MI.

Michie, K., S. Blackman and T. E. B. Ogunbiyi (1999). Twin-wire gmaw: process characteristics and applications. *Welding Journal* **78**(5), 31–34.

Misra, S., S. Subramanyam, S. Pandey and P. B. Sharma (1992). Expert system for GMA welding of aluminum. In: *Proceedings of ASM 3rd International Conference on Trends in Welding Research* (S. A. David and J. M. Vitek, Eds.). Gatlinburg, TN. pp. 949–956.

Moore, K. and D.S. Naidu (1997). Final report: Advanced welding project. Technical report. Idaho State University.

Moore, K., D. Naidu, R. Yender and J. Tyler (1997a). Gas metal arc welding control: Part i-modeling and analysis. In: *Nonlinear Analysis: Theory, Methods, and Applications.* pp. 3101–3111.

Moore, K., J. Tyler, S. Ozcelik and D.S. Naidu (1998a). Classical control of gas-metal arc welding. In: *American Control Conference.* Philadelphia, PA.

Moore, K. L. (1997). Iterative learning control - an expository review. Technical Report T.R. 96/97 003. Idaho State University. Pocatello, ID.

Moore, K. L. and A. Mathews (1997). Iterative learning control of systems with non-uniform trial length with applications to gas metal arc welding. In: *Proc. of the 2nd Asian Control Conference.* Seoul, Korea.

Moore, K. L., M. A. Abdelrahaman and D. S. Naidu (1999). Gas metal arc welding control - ii control strategy. *Nonlinear Analysis, Theory, Methods & Applications* **35**(1), 85–93.

Moore, K. L., M. A. Abdelrahman and D. S. Naidu (1997b). Gas metal arc welding control: Part ii - control strategy. *Nonlinear World.* (Accepted for publication).

Moore, K., M.A. Abdelrahman and D.S. Naidu (1998b). Gas metal arc welding control: Part ii-control strategy. In: *Nonlinear World Conference.*

Moore, K.L., D.S. Naidu, M. Abdelrahaman and A. Yesildirek (1996a). Advanced welding control project: Annual report FY96. Technical report. Measurement and Control Engineering Research Center. Idaho State University, Pocatello, ID.

Moore, K.L., D.S. Naidu, M.A. Abdelrahman and A. Yesildirek (1996*b*). Advanced welding control project: Annual report fy96. Technical report. Idaho State University.

Moore, K.L., D.S. Naidu, S. Ozcelik, B. Yender and J. Tyler (1997*c*). Advanced welding control project: Annual report fy97. Technical report. Idaho State University.

Moore, K.L., D.S. Naidu, S. Ozcelik, R. Yender and J. Tyler (1997*d*). Advanced welding control project: Annual report, FY97. Technical report. Measurement and Control Engineering Research Center. Idaho State University, Pocatello, ID.

Moore, K.L., M.A. Abdelrahman and D.S. Naidu (1996*c*). Gas metal arc welding control: part ii- control strategy. In: *Proceedings of the Second World Congress of Nonlinear Analysis*. Athens, Greece.

Moore, K.L., M.A. Abdelrahman and D.S. Naidu (1996*d*). Gas metal arc welding control: part ii- control strategy. In: *2nd World Congress of Nonlinear Analysis*. Athens, Greece.

Morgan, C. G., J. S. E. Bromley, P. G. Davey and A. R. Vidler (1983). Visual guidance techniques for robot arc welding. In: *Proceedings of SPIE Third International Conference on Robot Vision and Sensory Controls RoViSeC3*. Vol. 449. Cambridge, MA. pp. 390–399.

Mosca, E. (1995). *Optimal Predictive, and Adaptive Control*. Prentice Hall. Englewood Cliffs, NJ.

Murakami, E., K. Kugai and H. Yamamoto (1994). Dynamic analysis of arc length and its application to arc sensing. In: *Sensors and Control Systems in Arc Welding* (H. Nomura, Ed.). Chap. 25, pp. 216–227. Chapman & Hall. London, UK. English Translation of the Original 1991 Japanese Edition.

M.Wadsworth, H., Ed.) (1990). *Handbook of Statistical Methods for Engineers and Scientists*. McGraw Hill. New York, NY.

Nachev, G., B. Petkov and L. Blagoev (1983*a*). Data processing problems for gas metal arc GMA welder. In: *Proceedings of the SPIE 3rd International Conference on Robotic Vision and Sensory Controls RoViSEC3* (D. P. Casssssent and E. L. Hall, Eds.). Vol. 449. Cambridge, MA. pp. 291–296.

Nachev, G., B. Petkov, L. Blagoev and I. Tsankarski (1983b). Adaptive gas metal arc GMA welder. In: *Proceedings of th SPIE 3rd International Conference on Robot Vision and Sensory Controls, RoViSeC3*. Vol. 449. Cambridge, MA. pp. 286–290.

Nadeau, F., J. Blain and M. Dufour (1990). Computerized system automates GMA pipe welding. *Welding Journal* **69**, 375–385.

Nagarajan, S., K. N. Groom and B. A.Chin (1989a). Infrared sensors for seam tracking in gas tungsten arc welding processes. In: *Proceedings of the ASM 2nd International Conference on Trends in Welding Research* (S. A. David and J. M. Vitek, Eds.). Gatlinburg, TN. pp. 951–955.

Nagarajan, S., W. H. Chen and B. A. Chin (1989b). Infrared sensing for adaptive arc welding. *Welding Journal* **68**(11), 462s–466s.

Narendra, K. S. and Y. Lin (1980). Stable discrete adaptive control. *IEEE Transactions on Automatic Control*.

Nat (1994). *LabVIEW User Manual for Windows*.

Nayak, N. and A. Ray (1993). *Intelligent Seam Tracking for Robotic Welding*. Advances in Industrial Control. Springer-Verlag. London, UK.

Nayak, N., D. Thompson, A. Ray and A. Vavrek (1987). Conceptual development of an adaptive real-time seam tracker for welding automation. In: *Proc. of 1987 IEEE Intl. Conf. on Robotics and Automation*. Raleigh, NC. pp. 1019–1024.

Newton-Montiel, B. (1990). Workshop launches welding in space research. *welding Journal* pp. 45–47.

Niepold, R. and F. Brümmer (1984). A visual sensor for seam tracking and on-line process parameter control in arc-welding applications. In: *Proc. of 14th Intl. Symp. on Industrial Robots and 7th Intl. Conf. on Industrial Robot Technology*. pp. 375–385.

Nishar, D. V. (1992). Feedback control of bead temperature in gas metal arc welding. Technical report. University of Illinois. Urbana, IL.

Nomura, H. (1994a). Analysis of questionnaire results on sensor applications to welding processes. In: *Sensors and Control Systems in Arc Welding*. Chap. 6, pp. 76–85. Chapman & Hall. London, UK. English Translation of the Original 1991 Japanese Edition.

Nomura, H. (1994b). Basics for welding process control. In: *Sensors and Control Systems in Arc Welding*. Chap. 4, pp. 53–68. Chapman & Hall. London, UK. English Translation of the Original 1991 Japanese Edition.

Nomura, H. (1994c). Control systems. In: *Sensors and Control Systems in Arc Welding*. Chap. 3, pp. 44–52. Chapman & Hall. London, UK. English Translation of the Original 1991 Japanese Edition.

Nomura, H. (1994d). Future trends. In: *Sensors and Control Systems in Arc Welding*. Chap. 7, pp. 86–87. Chapman & Hall. London, UK. English Translation of the Original 1991 Japanese Edition.

Nomura, H. (1994e). Introduction to sensing systems and their application in arc welding processes. In: *Sensors and Control Systems in Arc Welding*. Chap. 1, pp. 1–17. Chapman & Hall. London, UK. English Translation of the Original 1991 Japanese Edition.

Nomura, H. (1994f). State-of-the-art review of sensors and their application in arc welding processes. In: *Sensors and Control Systems in Arc Welding*. Chap. 2, pp. 18–43. Chapman & Hall. London, UK. English Translation of the Original 1991 Japanese Edition.

Nomura, H., Ed.) (1994g). *Sensors and Control Systems in Arc Welding*. Chapman & Hall. London, UK. English Translation of the Original 1991 Japanese Edition.

Nomura, H., T. Yohida and K. Tohno (1976). Control of weld penetration. *Metal Construction* **8**, 244–246.

Norrish, J. (1988). Computer based instrumentation for welding. In: *Proceedings of Conference on Computer Technology in Welding*. The Welding Institute, Cambridge, UK.

Norrish, J. (1992). *Advanced welding processes*. Institute of Physics Publishing. Bristol, UK.

Oberly, P. A., M. G. D'Alillio, J. E.Jones, X. Xu and D. R. White (1989). Investigation of a blackboard artificial intelligence computer architecture for welding procedure specification and structural integrity analysis. In: *Proceedings of the ASM 2nd International Conference on Trends in Welding Research* (S. A. David and J. M. Vitek, Eds.). Gatlinburg, TN. pp. 985–989.

O'Brien, R. L., Ed.) (1991). *Welding handbook: Welding processes*. Vol. 2. Eighth ed.. American Welding Society. Maimi, FL.

Ogasawara, T., T. Maruyama, T. Saito, M. Sato and Y. Hida (1987). A power source for gas shielded arc welding with new current weld forms. *Welding Journal* pp. 57–.

Ogilvie, G. J. and I. M. Ogilvy (1983). The pulsed GMA process in automatic welding. In: *Proceedings of 31st Annual Conference of The Australian Welding Institute.* Sydney, Australia. pp. 16–19.

Ogilvie, K. S. (1991). Modeling of the gas metal arc welding process for control of arc length. PhD thesis. University of Michigan. Ann Arbor, MI. Dissertation.

Ohring, S. and H. J. Lugt (1999). Numerical simulation of a time-dependent 3-d gma weld pool due to a moving arc. *Welding Journal (Miami, Fla)* **78**(12), 416s–424s.

Ohshima, K., S. Yamane, H. Iida, S. Xiang, Y. Mori and T. Kobota (1994a). Fuzzy control of CO_2 short-arc welding. In: *Sensors and Control Systems in Arc Welding* (H. Nomura, Ed.). Chap. 15, pp. 137–146. Chapman & Hall. London, UK. English Translation of the Original 1991 Japanese Edition.

Ohshima, K., S. Yamane, Y. Mori and P. Ma (1994b). Application of a fuzzy neural network to welding line tracking. In: *Sensors and Control Systems in Arc Welding* (H. Nomura, Ed.). Chap. 16, pp. 147–1153. Chapman & Hall. London, UK. English Translation of the Original 1991 Japanese Edition.

Onda, H., Y. Nishinaga and K. Ono (1992). Welding defect identification by artificial neural networks. In: *Japan/USA Symposium on Flexible Automation.* pp. 313–316.

Oomen, G. L. and W. J. P. A. Verbeek (1983). A real-time optical profile sensor for robot arc welding. In: *Proceedings of Third International Conference on Robot Vision and Sensory Controls RoViseC3*. Vol. 449. Cambridge, MA. pp. 67–71.

Ortega, A. R., L. A. Bertram, E.A. Fuchs, K. W. Mahin and D. V. Nelson (1992). Thermomechanical modeling of a stationary gas metal arc weld: a comparison between numerical and experimental results. In: *Proceedings of the ASM 3rd International Conference on Trends in Welding Research* (S. A. David and J. M. Vitek, Eds.). Gatlinburg, TN. pp. 89–93.

Ouden, G., Y. H. Xiao and M. J. Hermans (1993). The role of weld pool oscillation in arc welding. *International Journal for the Joining of Materials* **5**(4), 123–129.

Ozcelik, S. (1996). Design of Robust Feed forward Compensators for Direct Model Reference Adaptive Controllers. PhD thesis. Rensselaer Polytechnic Institute. Troy, NY.

Ozcelik, S., K. L. Moore and D. S. Naidu (1997). Adaptive control of a gas metal arc welding (gmaw) process. Technical report. Measurement and Control Engineering Research Center, Idaho State University. Pocatello, ID.

Ozcelik, S., K. Moore and D.S. Naidu (1998*a*). Application of mimo direct adaptive control to gas metal arc welding. In: *American Control Conference*. Philadelphia, PA.

Ozcelik, S., K. Moore and D.S. Naidu (1998*b*). Mimo direct model reference adaptive control for gas metal arc welding. In: *The 5th Intrenational Conference on Trends in Welding Research*. Pine Mountain, GA.

P. Zhu, M. Rados and S. W. Simpson (1997). Theoretical predictions of the start-up phase in gma welding. *Welding Journal* **76**(7), 269s–274s.

Pan, J. L. (1986). Study of welding arc control in china. *Welding Journal* pp. 37–46.

Pan, J. L., R.H. Zhang, Z. M. Ou, Z. Q. Wu and Q. Chen (1989). Adaptive control GMA welding - a new technique for quality control. *Welding Journal* **68**(3), 73–76.

Pandey, S. and R. S. Parmar (1989). Mathematical models for predicting bead geometry and shape relationships for MIG welding of aluminum alloy 5083. In: *Proceedings of the ASM 2nd International Conference on Trends in Welding Research* (S. A. David and J. M. Vitek, Eds.). Gatlinburg, TN. pp. 37–41.

Papritan, J. C. and S. C. Helzer (1991). Statistical process control for welding. *Welding Journal* pp. 44–48.

Park, S. J. and T. J. Lim (1999). Fault-tolerant robust supervisor for discrete event systems with model uncertainty and its application to a workcell.. *IEEE Transactions on Robotics and Automation* **15**(2), 386–391.

Parshin, V. A., A. K. Seliverstov, A. V. Parfenova, S. B. Shakhanov and V. I. Shakhvatov (1981). Algorithm for the control of the dimensions of welded joints in thin sheets of high strength steel. *Welding Production (GB)* **28**(10), 3–5.

Pavone, V. J. (1983). Univision 11, a vision system for arc welding robots. In: *Proceedings of AWS Conference on Automation and Robotics For Welding*. pp. 91–103.

Payares, C. M., M. Dorta and P. E. Munoz (1999). Influence of the welding variables on the mechanical properties in butt joints for aluminum 6063-t5. *American Society of Mechanical Engineers. Pressure Vessels and Piping Division* **393**, 339–344.

Peters, C. N. (1986). The use of backface penetration control methods in synergetic pulsed MIG welding. PhD thesis. Cranfield Institute of Technology. Cranfield, UK.

Peterson, C. (1987). *Statistical Process Control*. Viking Press Inc.. Waterloo, Iowa.

Pezzutti, M. (2000). Innovative welding technologies for the automotive industry. *Welding Journal* **79**(6), 43–46.

BIBLIOGRAPHY

Phillips, C. L. and H. T. Nagle (1995). *Digital Control System Analysis and Design, Third Edition.* Prentice Hall. Englewood Cliffs, NJ.

Phillips, C. L. and R. D. Harbor (1996). *Feedback Control Systems, Third Edition.* Prentice Hall. Englewood Cliffs, NJ.

Placko, D., H. Clergeot and F. Monteil (1985). Seam tracking using a linear array of eddy current sensors. In: *Proceedings of the 5th Intl. Conf. on Robotic Vision and Sensory Controls* (N. J. Zimmerman, Ed.). Amsterdam, The Netherlands. pp. 557–568.

Postacioglu, N., P. Kapadia and J. Dowden (1991). Theory of the oscillations of an ellipsoidal weld pool in laser welding. *Journal Physics D: Applied Physics* **24**, 1288–1292.

Quimby, B. J. and G. D. Ulrich (1999). Fume formation rates in gas metal arc welding. *Welding Journal (Miami, Fla)* **78**(4), 142s–149s.

Quinn, T. P. and R. B. Madigan (1992). Dynamic model of electrode extension for gas metal arc welding. In: *Proceedings of ASM 3rd International Conference on Trends in Welding Research* (S. A. David and J. M. Vitek, Eds.). Gatlinburg, TN. pp. 1003–1008.

Quinn, T. P. and R. B. Madigan (1993). Adaptive arc length controller design for GMAW. In: *Proceedings of American Welding Society International Conference on Modeling and Control of Joining Processes.*

Quinn, T. P., C. Smith, C. N. McCowan, E. Blackhowaik and R. B. Madigan (1999). Arc sensing for defects in constant-voltage gas metal arc welding. *Welding Journal* pp. 322s–328s.

Quinn, T. P., R. B. Madigan and T. A. Siewert (1994). An electrode extension model for gas metal arc welding. *Welding Journal* **73**, 241s–247s.

Quintino, L. and C. J. Allum (1984a). Pulsed GMAW: interactions between process parameters - part 1. *Welding and Metal Fabrication* pp. 85–89.

Quintino, L. and C. J. Allum (1984b). Pulsed GMAW: interactions between process parameters - part 2. *Welding and Metal Fabrication* pp. 126–129.

Quintino, L., R. Riberio and J. Faira (1999). Classification of gmaw transfer using neural networks. *International Journal for the Joining of Materials* **11**(1), 6–8.

Rajasekaran, S. (1999). Weld bead characteristics in pulsed gma welding of al-mg alloys. *Welding Journal (Miami, Fla)* **78**(12), 397s–407s.

Rajasekaran, S., S. D. Kulkarni, U. D. Mallya and R. C. Chaturvedi (1994). Molten droplet detachment characteristics in steady and pulsed current GMAwelding Al-Mg alloys. Technical report. The Weld Institute Research Report.

Rajasekaran, S., S. D. Kulkarni, U. D. Mallya and R. C. Chaturvedi (1998). Droplet detachment and plate fusion charcteristics in pulsed current gas metal arc welding. *Welding Journal* **77**(6), 254s–269s.

Ramaswamy, K., K. Andersen and G. E. Cook (1989). New techniques for modeling and control of GTA welding. In: *Proc. of IEEE Southeastcon '89*. Columbia, SC. pp. 1250–1260.

Ramsey, P. W., J. J. Chyle, J. N. Kuhr, P. S. Myers, M. Weiss and W. Groth (1963). Infrared temperature sensing systems for automatic fusion welding. *Welding Journal* **42**(8), 337s–346s.

Randhawa, H. S., P. K. Ghosh and S. R. Gupta (2000). Some basics aspects of geometrical characteristics of pulsed current vertical-up gma weld. *ISIJ International* **40**(1), 71–76.

Ray, S., M. Bhattacharyya and S. N. Banerjee (1986). Investigation into optimal welding variables for arc welded butt joints. *Welding Journal* pp. 39s–44s.

Reeves, R. E., T. D. Manley, A. Potter and D. R. Ford (1988). Expert system technology -an avenue to an intelligent weld process control system. *Welding Journal* pp. 33–41.

Reilly, R. (1991). Real-time weld quality monitor control GMAwelding. *Welding Journal* pp. 36–41.

Reutzel, E. W., C.J. Einerson, J. A. Johnson, H. B. Smartt, T. Harmer and K. L. Moore (1995a). Derivation and calibration of a gas metal arc welding GMAWdynamic droplet model. In: *Proceedings of the ASM 4rd International Conference on Trends in Welding Research* (H. B. Smartt, J. A. Johnson and S. A. David, Eds.). Gatlinburg, TN. pp. 377–384.

Reutzel, E.W., C. Einerson, J. Johnson, H Smartt, T Harmer and K. Moore (1995b). Derivation and calibration of a gas metal arc welding dynamic droplet model. In: *the 4th International Conference on Trends in Welding Research*. Gatlinburg, TN.

Rhode, D. and P. V. Kokotovic (1989). Parameter convergence conditions independent of plant order. In: *Proc. of 1989 American Control Conference*. Pittsburgh, PA. pp. 981–986.

Richardson, R. A., D. A. Gutow and S. H. Rao (1982). A vision based system for arc weld pool size control. In: *Measurement and Control for Batch Manufacturing* (D. E. Hardt, Ed.). pp. 65–75. ASME. New York, NY.

Richardson, R. D. and R. W. Richardson (1983). The measurement of two-dimensional arc weld pool geometry by image analysis. In: *Proceedings of the ASME Winter Annual Meeting on Control of Manufacturing Processes and Robotic Systems* (D. E. Hardt and W. J. Book, Eds.). ASME. Boston, MA. pp. 137–148.

Richardson, R. W. (1981). Review of the state-of-the-art of adaptive control for the gas tungsten and plasma arc welding processes. Technical report. Air Force Materials Laboratory, Wright-Patterson Air Force Base. Dayton, OH.

Richardson, R. W., A. Gutow, R. A. Anderson and D. F. Farson (1984). Coaxial arc weld pool viewing for process monitoring and control. *Welding Journal* pp. 43–50.

Richardson, R. W. and C. C. Conrardy (1992). Coaxial vison-based control of GMAW. In: *Proceedings of ASM 3rd International Conference on Trends in Welding Research* (S. A. David and J. M. Vitek, Eds.). Gatlinburg, TN. pp. 957–961.

Richardson, R. W. and R. A. Anderson (1983). Weld butt joint tracking with a coaxial viewer based weld vision system. In: *Proceedings of ASME Winer Annual Meeting on Control of Manufacturing Process and Robotic Systems* (D. E. Hardt and W. J. Book, Eds.). ASME. Boston, MA. pp. 107–119.

Richardson, R. W., w. A. Penix and K. L. Boyer (1992). Interpretation of arc weld images by vision analysis. In: *Japan/USA Symposium on Flexible Automation.* pp. 309–312.

Rider, G. (1979). On line measurement of weld pool surface size. In: *Proceedings of Welding and Fabrication in the Nuclear Industry.* British Nuclear Energy Society. London, UK. pp. 351–359.

Rider, G. (1983). Control of weld pool size and position for automatic and robotic welding. In: *Proceedings of the SPIE Third International Conference on Robot Vision and Sensory Controls RoViseC3.* Vol. 449. Cambridge, MA. pp. 381–389.

Rioux, M. (1984). Laser range finder based on synchronized scanners. *Applied Optics.*

Robinson, H. and N. R. Nutter (1980). The development of automatic welding techniques for nuclear applications. In: *Proceedings of an International Conference on Developments in Mechanized Automated and Robotic Welding.* London, UK. pp. P31–P315.

Rock, A., X. Xu and J. E. Jones (1989). Investigation of an artificial neural system for a computerized welding vision system. In: *Proceedings of the ASM 2nd International Conference on Trends in Welding Research* (S. A. David and J. M. Vitek, Eds.). Gatlinburg, TN. pp. 957–965.

Rokhlin, S. I. (1989). In-process radiographic evaluation of arc welding. *Materials Evaluation* **47**, 219–224.

Rokhlin, S. I. and A. C. Guu (1990). Computerized radiographic sensing and control of an arc welding pro-cess. *Welding Journal* pp. 83s–97s.

Rokhlin, S. I., K. Cho and A. C. Guu (1989). Closed-loop process control of weld pool penetration using real-time radiography. *Materials Evaluation* **47**, 363–369.

Romanenkov, E. I. (1976). Control of arc length on the basis of its spectral radiation. *Welding Production* **23**, 53–54.

Rosenthal, D. (1941). Mathematical theory of heat distribution during welding and cutting. *Welding Journal* **20**(5), 220s–234s.

Rosenthal, D. (1946). The theory of moving surfaces of heat and its application to metal treatment. *Transactions of ASME* pp. 849–866.

S. M. Govardhan, H. C. Wikle, S. Nagarajan and B. A. Chin (1995). Real-time welding process control using infrared sensing. In: *Proceedings of the American Control Conference.* Seattle, WA. pp. 1712–1716.

Saedi, H. R. and W. Unkel (1988). Arc and weld pool behavior for pulsed current GTAW. *Welding Journal* **67**(11), 247s–255s.

Salter, R. J. and R. T. Deam (1986). A practical front face penetration control system for TIG welding. In: *Proceedings of the Conference on Developments in Automated and Robotic Welding.* The Welding Institute, London, UK.

Sampson, R. E., G. Suits and C. B. Arnold (1985). Analysis of sensors for application to welding control. Technical Report N00167-84-K-0036. University of Michigan. Ann Arbor, MI.

Santos, T. O., R. B. Caetano, J. M. Lemos and F. J. Coito (2000). Multipredictive adaptive control of arc welding trailing centerline temperature. *IEEE Transactions on control Systems Technology* **8**(1), 159–169.

Schiano, J. L. (1993). Feedback control of two physical processes: design and experiments. Technical Report DC-148, UILU-ENG-93-2217. University of Illinois. Urbana, IL.

Schick, W. R. (1988). Verticle strip cladding: process control. *welding Journal* pp. 17–22.

Schrier, H., M. Sutton, Y. Chao, H. Bruck and J. Dydo (1999). Full-field temperature and three-dimensional displacement measurements in hostile environments. *Society of Manufacturing Engineers* **MR99-140**, 1–5.

Schulkes, R.M. (1994). The evolution and bifurcation of a pendant drop. *Journal of Fluid Mechanics* **278**, 83–100.

Seborg, D.E., T.F. Edgar and D.A. Melichamp (1992). *Process Dynamics and Control*. John Wiley and Sons. New York, NY.

Segatskii, G. I., S. V. Dubovetskii and O. G. Kasatkin (1983). Models for open control of CO_2 weld formation. *Automatic Welding(GB)* **36**(2), 18–21.

Shaw, R. (1984). *The Dripping Faucet as a Model Chaotic System*. Ariel Pess, Inc.,. Santa Cruz, CA.

Shepard, M. E. (1991). Modeling of Self-Regulation in Gas Metal Arc welding. PhD thesis. Vanderbilt University. Nashville, TN.

Shepard, M. E. and G. E. Cook (1992). A frequency-domain model of self-regulation in gas metal arc welding. In: *Proceedings of the ASM 3rd International Conference on Trends in Welding Research* (S. A. David and J. M. Vitek, Eds.). Gatlinburg, TN. pp. 899–903.

sicard, P. and M. D. Levine (1987). Automatic joint recognition and tracking for robotic arc welding. In: *Proceedings of IEEE Montech '87-Compint'87*. Montreal, Canada. pp. 290–293.

Sicard, P. and M.D. Levine (1988). An approach to an expert robot welding system. *IEEE Transactions on Systems, Man and Cybernetics* **18**(2), 204–222.

Siewert, T. A., R. B. Madigan and T. P Quinn (1997). Sensors control gas metal arc welding. *Advanced Materials & Processes* **151**(4), 23–25.

Siewert, T. A., R. B. Madigan, T. P. Quinn and M. A. Mornis (1992*a*). Through-the-arc sensing for monitoring arc welding. In: *Proceedings of the ASM 3rd International Conference on Trends in Welding Research* (S. A. David and J. M. Vitek, Eds.). Gatlinburg, TN. pp. 1037–1040.

Siewert, T. A., R. B. Madigan, T. P. Quinn and M. A.Mornis (1992*b*). Through-the-arc sensing for real-time measurement of gas metal arc weld quality. In: *In Proceedings of International Conference* on computerization of Welding Information IV, Orlando, FL.. National Inst. of Standards and Technology MSEL. Gaithersburg, MD. pp. 198–206.

Slotine, J.-J. and W. Li (1991). *Applied Nonlinear Control*. Prentice Hall. Englewood Cliffs, NJ.

Smartt, H. B. (1992). Intelligent sensing and control of arc welding. In: *Proceedings of the ASM 3rd International Conference on Trends in Welding Research* (S. A. David and J. M. Vitek, Eds.). Gatlinburg, TN. pp. 843–851. Keynote Address.

Smartt, H. B., A. D. Watkins and M. D. Light (1989a). Sensing and control problems in gas metal welding. In: *Proceedings of the ASM 2nd International Conference on Trends in Welding Research* (S. A. David and J. M. Vitek, Eds.). Gatlinburg, TN. pp. 917–921.

Smartt, H. B. and C. J. Einerson (1993a). A model for heat and mass input control in gas metal arc welding. *Welding Journal* **72**(5), 217s–229s.

Smartt, H. B. and C. J. Einerson (1993b). A model for heat and mass input control in gas metal arc welding. *Welding Journal* pp. 217–229.

Smartt, H. B. and J. A. Johnson (1989). Intelligent control of arc welding. Technical report. EG&G, Idaho National Engineering Laboratory. Idaho Falls, ID. Preprint.

Smartt, H. B. and J. A. Johnson (1991). A novel implementation of fuzzy logic – theory. Unpublished technical report. EG&G, Idaho, Idaho National Engineering Laboratory. Idaho Falls, ID.

Smartt, H. B., C. J. Einerson, A. D. Watkins and R. A. Morris (1986). Gas metal arc process sensing and control. In: *Proceedings of an International Conference on Trends in Welding Research*. ASM. Gatlinburg, TN. pp. 461–465.

Smartt, H. B., C. J. Einerson and A. D. Watkins (1988). Computer control of gas metal arc welding. In: *Computer technology in Welding, Cambridge, UK*. Cambridge, UK.

Smartt, H. B., C. J. Einerson and A. D.watkins (1989*b*). Methods for controlling gas metal arc welding. Patent. EG&G Idaho, Inc., Idaho falls, ID.

Smartt, H. B., J. A. Johnson, C. J. Einerson, A. D. Watkins and N. M. Carlson (1990). Model-based approach to intelligent control of gas metal arc welding. In: *Proceedings of the 5th International Conference on Modeling of Casting, Welding and Advanced Solidification Processes* (M. Rappaz, M. R. Ozgu and K. W. Mahin, Eds.). Davos, Switzerland. pp. 305–313.

Smartt, H. B., Johnson, J. A. and David, S. A., Eds.) (1995). *Proceedings of the ASM 4rd International Conference on Trends in Welding Research*. ASM International. Gatlinburg, TN.

Smati, Z., D. Yapp and C. J. Smith (1984). Laser guidance system for robotics. In: *Proceedings of 4th International Conference on Robot Vision and Sensory Controls RoViseC4*. London, UK. pp. 91–101.

Smati, Z., P. J. Alberry and D. Yapp (1987). Strategies for automatic multipass welding. In: *Proceedings of First International Conference on Advanced Welding System* (P. T. Houldcroft, Ed.). Vol. 519. London, UK. pp. 219–237.

Solomon, H. D. and S. Levy (1982). HAZ temperatures and cooling rate as determined by a simple computer program. In: *Trends In Welding Research In The United States* (S. A. David, Ed.). pp. 173–205. American Society for Metals. Metals Park, OH.

Song, J. B. and D. E. Hardt (1990). Development of a heat-transfer based depth estimator for real-time welding control. In: *Proceedings of ASME, Symposium on Manufacturing Process Modeling and Control.*

Song, J. B. and D. E. Hardt (1991). Multivariable adaptive control of bead geometry in GMA welding. In: *Proceedings of ASME Winter Annual Meeting of the American Society of Mechanical Engineers*. Atlanta, GA.

Song, J. B. and D. E. Hardt (1992*a*). Simultaneous control of bead width and depth geometry in gas- metal arc welding. In: *Proceedings of Third International Conference on Welding Research, ASME*. Gatlinburg, TN.

Song, J. B. and D. E. Hardt (1992b). A thermally based weld pool depth estimator for real-time control. In: *Proceedings of 3rd International Conference on Trends in welding research.* Gatlinburg, TN. pp. 975–980.

Song, J. B. and D. E. Hardt (1993). Application of adaptive control to arc welding process. In: *Proceedings of the American Control Conference.* San Francisco, CA. pp. 1751–1755.

Song, J. B. and D. E. Hardt (1994). Dynamic modeling and adaptive control of the gas metal arc welding process. *Transactions of the ASME, Journal of Dynamic Systems, Measurement and Control* **116**(3), 405–413.

Sorensen, C. D. and T. W. Eagar (1990). Modeling of oscillation in partially penetrated weld pools. *Transactions of the ASME, Journal of Dynamic Systems, Measurement and control* **112**(3), 469–474.

Spraragen, W. and B. A. Lengyel (1943). Physics of the arc and the transfer of metal in arc welding: A review of the literature to February 1942. *Weldig Journal* **1**, 2s–42s.

stefan, A., B. Gunnar B. Ali and C. Ingvar (1998). Quality monitoring in robotized spray gma weld. *International Journal for the Joining of Materials* **10**(1-2), 3–23.

Stefanuk, W. P., J. N. Rempel, J. P. Huissoon and H. W. Kerr (1992). A multivariable approach to the control of fillet weld dimensions. In: *Proceedings of the ASM 3rd International Conference on Trends in Welding Research* (S. A. David and J. M. Vitek, Eds.). Gatlinburg, TN. pp. 889–893.

Stenbacka, N. and O. Svensson (1987). Some observations on pore formation in gas metal arc welding. *Scandinavian journal of Metallurgy* **16**(4), 151–153.

Stenke, V. (1990). Economic aspects of active gas metal arc welding. *Svetsaren, a welding review* **xx**(2), 11–13.

Sthen, T. and T. Porsander (1983). An adaptive torch position system for welding of car bodies. In: *Proc. of the 3rd Intl. Conf. on Robot Vision and Sensory Controls.* Cambridge, MA. pp. 607–613.

Stol, I. (1985). Advanced gas metal arc welding process GMAW. In: *Proceedings of the First EWI International Conference on Advanced Welding System* (P. T. Houldcroft, Ed.). Vol. 519. Edison Welding Institute. London, UK. pp. 493–511.

Stol, I. (1989). Development of an advanced gas metal arc welding process. *Welding Journal* **68**(8), 313s–326s.

Stone, D., J. S. Smith and J. Lucas (1990). Sensor for automated weld bead penetration control. *Measurement Science and Technology* **1**, 1143–1148.

Stroud, R. R. (1989). Problems and observations whilst dynamically monitoring molten weld pools using ultra sound. *British Journal of Non-Destructive Testing* **31**, 29–32.

Stroud, R. R. and R. Fenn (1983). Microcomputer control of weld penetration. In: *Proceedings of 31st Annual Conference on Welding and Computers*. Sydney, Australia. pp. 10–14.

Stroud, R. R. and T. J. Harris (1990). Seam tracking butt and filler welds using ultrasound. *Joining of Materials*.

Subramaniam, S., D. R. White, D. J. Scholl and W. H. Weber (1998). In situ optical measurements of liquid drop tension in gas metal arc welding. *Journal of Physics D: Applied Physics* **31**(16), 1963–1967.

Subramaniam, S., D.R. White, J.E. Jones and D.W. Lyons (1999). Experimental approach to selection of pulsing parameters in pulsed gmaw. *Welding Journal (Miami, Fla)* **78**(5), 166s–172s.

Sugitani, Y. and W. Mao (1996). Automatic simultaneous control of bead height and back bead shape using an arc sensor in one-sided welding with a backing plate. *Welding Research Abroad* **42**(12), 9–17.

sugitani, Y., Y. Kanjo and M. Ushio (1999). High efficciency processes in automatic welding. *Welding Research Abroad* **45**(4), 32–37.

Sun, J. S., C. S. Wu and J. Q. Gao (1999). Modeling the weld pool behaviors in gma welding. *International Journal for the Joining Materials* **11**(4), 112–117.

Suzuki, A. and D. E. Hardt (1987). Application of adaptive control theory to in-process weld geometry regulation. In: *Proceedings of the 1987 American Control Conference.* pp. 723–728.

Suzuki, A. and D. E. Hardt (1988). Application of adaptive control theory to on-line GTA weld geometry regulation. In: *Proceedings of the Winter Annual Meeting of the ASME on Control Methods for the Manufacturing Processes* (D. E. Hardt, Ed.). Chicago, IL. pp. 13–25.

Suzuki, A., D. E. Hardt and L. Valavani (1991). Application of adaptive control theory to on-line GTA weld geometry regulation. *Transactions of the ASME Journal of Dynamic Systems, Measurement and Control* **113**, 93–103.

Sweet, L. M. (1985). Sensor-based control systems for arc welding robots. *Robotics and Computer-Integrated Manufacturing* **2**(2), 125–133.

Sweet, L. M., A. W. Case Jr, N. R. Corby and N. R. Kuchar (1983). Closed-loop joint tracking, puddle centering and weld process control using an integrated weld torch vision system. In: *Proceedings of the ASME Winter Annual Meeting on Control of Manufacturing Processes and Robotic Systems* (D. E. Hardt and W. J. Book, Eds.). Boston, MA. pp. 97–105.

Tade, M. O. and D. W. Bacon (1986). Adaptive decoupling of a class of multivariable dynamic systems using output feedback. *IEE Proceedings.*

Tam, A. S. and D. E. Hardt (1989). Weld pool impedance for pool geometry measurement: stationary and nonstationary pools. *ASME Transactions: Journal of Dynamic systems, Measurement and Control* **111**, 545–553.

Tan, C. and J. Lucas (1986). Low cost sensors for seam tracking in arc welding. In: *Proceedings of the 1st International Conference on Computer Technology in Welding.* The Welding Institute, London, UK.

Tandon, R. K., J. Ellis, P.T. Crisp and R. S. Baker (1984). Fume generation and melting rates of shielded metal arc welding electrodes. *Welding Journal* pp. 263s–266s.

Tao, J. and P. Levick (1999). Assessment of feedback variables for through the arc seam tracking in robotic gas metal arc welding. In: *Proceedings of Feedback Variable for through the arc seam tracking in robotic gas metal arc welding*. Pscataway, New Jersey. pp. 3050–3052.

Taylor, P. L., A. D. Watkins, E. D. Larsen and H. B. Smartt (1992). Integrated optical sensor for GMAW feedback control. In: *ASM 3rd International conference on Trends in Welding Research* (S. A. David and J. M. Vitek, Eds.). Gatlinburg, TN. pp. 1049–1053.

Tomizuka, M. (1988). Design of digital tracking controller for manufacturing applications. In: *Proceedings of the Winter Annual Meeting of the ASME on Control Methods for the Manufacturing Processes* (D. E. Hardt, Ed.). Chicago, IL. pp. 71–78.

Tomizuka, M., D. Dornfeld and M. Purcell (1980). Application of microcomputers to automatic weld quality control. *Transactions of the ASME: Journal of Dynamic Systems, Measurement and Control* **102**(2), 62–68.

Trindade, E. M. and C. J. Allum (1984). Characteristics in steady and pulsed current GMAW. *Welding and Metal Fabrication* pp. 264–272.

Tsai, C. L. (1982). Modeling of thermal behaviors of metals during welding. In: *Trends In Welding Research in the United States* (S. A. David, Ed.). pp. 91–108. American Society for Metals.

Tsai, C. L. (1983). Heat Distribution and Weld Bead Geometry in Arc welding. PhD thesis. MIT. Cambridge, MA.

Tsai, M. C. and S. Kou (1990). Electromagnetic -force- induced convection in weld pools with a free surface. *Welding Journal* pp. 241s–246s.

Tsukamoto, F., T. Hinata, K. Yasude and T. Onzawa (1999). Study on welding procedure of gas metal arc welding using cored wire. *Welding Research Abroad* **45**(11), 41–49.

Tusek, J. (1999). Mathematical model for the melting rate in welding with a multiple-wire electrode. *Journal of Physics D: Applied Physics* **32**(14), 1739–1744.

BIBLIOGRAPHY

Tyler, J. (1997a). Model-based control of gas metal arc welding process. Master's thesis. Idaho State University. Pocatello, ID.

Tyler, J. (1997b). Model-based control of the gas metal arc welding process. Master's thesis. Idaho State University. Pocatello, ID.

Tzafestas, G. S. and J. E. Kyriannakis (2000). Regulation of gma welding thermal characteristics via a hierarchical mimo predictive control scheme assuring stability. *IEEE Transactions on Industrial Electronics* **47**(3), 668–678.

Tzafestas, S. G., G. G. Rigatos and E. J. Kyriannakis (1997). Geometry and thermal regulation of gma welding via conventional and neural adaptive control. *Journal of Intelligent and Robotic Systems* **19**(2), 153–186.

Ueguri, S., K. Hara and H. Komura (1985). Study of metal transfer in pulsed GMA welding. *Welding Journal* pp. 242s–250s.

Ushio, M. and C. S. Wu (1997). Mathematical modeling of three-dimensional heat and fluid in a moving gas metal arc weld pool. *Metallurgical and Materials Transactions. B: Process Metallurgy and Materials Processing Science* **28**(3), 509–516.

Verdelho, P., M. Pio Silva, E. Margato and J. Esteves (1998). Electronic welder control circuit. In: *IECON Proceedings (Industrial Electronics Conference), Proceedings of the 1998 24th Annual Conference of the IEEE Industrial Electronics Society*. Los Alamitos, California. pp. 612–617.

Verdon, D. C., D. Langley and M. H. Moscardi (1980). Adaptive welding control using video signal processing. In: *Proceedings of the International Conference on Developments in Mechanized, Automated and Robot Welding*. London, UK. pp. P291–P2911.

Villafuerte, J. (1999). Understanding contact tip longevity for gas metal arc welding. *Welding Journal (Miami, Fla)* **78**(12), 29–35.

Vorman, A. R. and H. Brandt (1976). Feedback control of GTA welding using puddle width measurement. *Welding Journal* **55**(9), 742–749.

W. J. Kerth, Sr. and Jr. W. J. Kerth (1983). Digitally-controlled adaptive welding system. In: *Proceedings of Tenth NSF Conference on Production Research Technology*. pp. 139–140.

Waddoups, M. (1994). Detection and modeling of droplet detachment in gas metal arc welding. Master's thesis. Idaho State University. Pocatello, ID.

Wahab, M. A. and M. J. Painter (1996). Measurement and prediction of weld pool shape during gas metal arc welding using a noncontact laser profiling system. In: *Proceedings of SPIE - The International Society for Optical Engineering*. Singapore. pp. 34–39.

Wahab, M. A. and M. J. Painter (1997). Numerical model of gas metal arc welds using experimentally determined weld pool shapes as the representation of the welding heat source.. *International Journal of Pressure vessels and Piping* **73**(2), 153–159.

Wahab, M. A., M. J. Painter and M. H. Davies (1998). Prediction of the temperature distribution and weld pool geometry in the gas metal arc welding process.. *Journal of Materials Processing Technology* **77**(1-3), 233–239.

Wang, L.-X. (1997). *A Course in Fuzzy Systems and Control*. Prentice Hall PTR. Upper Saddle River, NY.

Waszink, J. H. and G. J. P. M. van den Heuvel (1979). Measurements and calculations of the resistance of the wire extension in arc welding. In: *Proceedings of the International Conference on Arc Physics and Weld pool Behavior*. London, UK. pp. 227–239.

Waszink, J. H. and G. J. P. M. van den Heuvel (1982). Heat generation and heat flow in the filler metal in GMA welding. *Welding Journal* **61**, 269s–282s.

Waszink, J. H. and L. H. Graat (1983). Experimental investigation of the forces acting on a drop of weld metal. *Welding Journal* **62**, 108s–116s.

Waszink, J. H. and M. J. Piena (1986). Experimental investigation of drop detachment and drop velocity in GMAW. *Welding Journal* pp. 289s–298s.

Watanabe, H., Y. Kondo and K. Inoue (1994). Automatic control technique for narrow-gap GMA welding. In: *Sensors and Control Systems in Arc Welding* (H. Nomura, Ed.). Chap. 21, pp. 182–190. Chapman & Hall. London, UK. English Translation of the Original 1991 Japanese Edition.

Watkins, A. D. (1989). Heat transfer efficiency in gas metal arc welding. Master's thesis. University of Idaho. Moscow, ID.

Watkins, A. D., H. B. Smartt and J. A. Johnson (1992). A dynamic model of droplet growth and detachment in GMAW. In: *Proceedings of ASM 3rd International Conference on Trends in Welding Research* (S. A. David and J. M. Vitek, Eds.). Gatlinburg, TN. pp. 993–1002.

Wegrzyn, T. (1999). Classification of metal weld deposits in terms of the amount of oxygen. In: *Proceediongs of the International Offshore and Polar Engineering Conference*. Golden, Colorado. pp. 212–216.

Wel (1980). *On-line control of the arc welding process*. London, UK.

Wezenbeek, H. C. (1992). A System for Measurement and Control of Weld Pool Geometry in Automatic Arc Welding. PhD thesis. Technische Univ. Eindhoven Netherlands, Dept. of Electrical Engineering.. Eindhoven, The Netherlands.

White, D. R. and J. E. Jones (1989). A hybrid hierarchical controller for intelligent control of welding processes. In: *Proceedings of the ASM 2nd International Conference on Trends in Welding Research* (S. A. David and J. M. Vitek, Eds.). Gatlinburg, TN. pp. 909–915.

White, D. R., J. A. Carmein, J. E. Jones and K. Liu (1992). Integration of process and control models for intelligent control of welding. In: *Proceedings of the ASM 3rd International Conference on Trends in Welding Research* (S. A. David and J. M. Vitek, Eds.). Gatlinburg, TN. pp. 883–887.

Wolovich, W. A. (1974). *Linear Multivariable Systems*. Springer-Verlag. New York, NY.

Won, Y. J. and H. S. Cho (1992). A fuzzy rule-based method for seeking stable arc condition under short-circuiting mode of GMA welding process. *Proceeding of the Institution of Mechanical Engineers,Part1: Journal of Systems and control Engineering* **206**, 117–125.

Won, Y. J. and H. S. Cho (1993). Fuzzy predictive approach to the control of weld pool size in gas metal arc welding processes. In: *Proceedings of ASME Winter Annual Meeting on Manufacturing Science and Engineering American Society of Mechanical Engi..* Vol. 64. ASME. New Orleans, LA. pp. 927–938.

Wonham, W. M. (1979). *Linear Multivariable Control: A Geometric Approach, Third Edition*. Springer-Verlag. New York, NY.

Wu, C. S. and J. S. Sun (1998). Modelling the arc heat flux distribution in gma welding. *Computational Materials Science* **9**(3-4), 397–402.

Wu, C. S., T. Polite and D. Rehfeldt (2001). A fuzzy logic system for process monitoring and quality evaluation in gmaw. *Welding Journal* pp. 33s–38s.

Wu, G.-D. and R. W. Richardson (1989). The dynamic response of self-regulation of the welding arc. In: *Proceedings of the ASM 2nd International Conference on Recent Trends in Welding Research* (S. A. David and J. M. Vitek, Eds.). Gatlinburg, TN. pp. 929–933.

Xiao, Y. H. and G. den Quden (1990). A study of GTA weld pool oscillation. *Welding Journal* **69**(8), 289s–293s.

Yamamoto, H., S. Harada and T. Ueyama (1997). Improved current control makes inverters the power sources of choice. *Welding Journal*.

Yamamoto, M., Y. Kaneko, K. Fujii, T. Kumazawa, K. Ohishima, G. Alzamora, T. Kubota, F. Ozaki and S. Anzai (1988). Adaptive control of pulsed MIG welding using image processing system. In: *Rec. of Conf. of the 23rd IEEE Industry Applications Society*. Pittsburgh, PA. pp. 1381–1386.

Yang, Z. and T. Debroy (1999). Modeling macro-microstructurs of gas-metal-arc welded hsla-100 steel. *Metallurgical and Materi-*

als *Transactions A: Physical Metallurgy and Materials Science* **30**(3), 483–493.

Yender, R.F. (1997). Design, construction, and modeling of an automated gas metal arc welding facility for controller research. Master's thesis. Idaho State University. Pocatello, ID.

Yetukuri, N. and G. W. Fischer (1997). Planning the gmaw process by constraint propagation. *Journal of Intelligent Manufacturing* **8**(6), 477–488.

Yoo, C. D. (1990). Effects of weld pool condition on pool oscillation. PhD thesis. The Ohio State University.

Yoo, C. D. and R. W. Richardson (1993). An experimental study on sensitivity and signal characteristics of weld pool and oscillation. *Transaction of Japan Welding Society* **24**, 54–62.

Yoo, C. D., K. I. Koh and H. K. Sunwoo (1997a). Investigation on arc light intensity in gas metal arc welding. part 2: Application to weld seam tracking. In: *Proceedings of the Institution of Mechanical Engineers.*. London, England. pp. 355–363.

Yoo, C. D., Y. S. Yoo and H. K. Sunwoo (1997b). Investigation on arc light intensity in gas metal arc welding. part1 : Relationship between arc light intensity and arc length. In: *Proceedings of the Institution Of Mechanical Engineers*. London, England. pp. 345–353.

Y.S.Kim and T. W. Eagar (1989). Temperature distribution and energy balance in the electrode during GMAW. In: *Proceedings of the ASM 2nd International Conference on Trends in Welding Research* (S.A. Divid and J. M. Vitek, Eds.). Gatlinburg, TN. pp. 13–18.

Zacksenhouse, M. and D. E. Hardt (1983). Weld pool impedance identification for size measurement and control. *ASME Transactions: Journal of Dynamic System, Measurement, and Control* **105**(9), 179–184.

Zadeh, L. A. (1965). Fuzzy sets. *Information and Control* **8**, 338–353.

Zhang, S. B., Y. M. Zhang and Kovacevic R. (1998a). Noncontact ultrasonic sensing for seam tracking in arc welding processes. *Journal of Manufacturing Science and Engineering, Transactions of the ASME.* **120**(3), 600–608.

Zhang, Y. M. and E. Liguo (2000). Numerical analysis of dynamic growth of droplet in gmaw. *Journal of Mechanical Engineering Science* pp. 1247–1258.

Zhang, Y. M. and P. J. Li (2001). Modified active control of metal transfer and pulsed gmaw of titanium. *Welding Journal* pp. 54s–61s.

Zhang, Y. M. and R. Kovacevic (1995). Modeling and real-time identification of weld pool characteristics for intelligent control. In: *The First Congress on Intelligent Manufacturing Process & Systems, CIRP.* Mayaguez, Puerto Roco.

Zhang, Y. M. and R. Kovacevic (1996). Monitoring of three-dimensional arc weld surface. In: *Accepted for the 26th Conference on Production Engineering.* Budva, Yugoslavia.

Zhang, Y. M. and R. Kovacevic (1997). Real-time sensing of sag geometry during GTA welding. *ASME Journal of Manufacturing Science and Engineering* **119**(2), 1–10.

Zhang, Y. M., B. L. Walcott and L. Wu (1992a). Adaptive predictive decoupling control of full penetration process in GTAW. In: *Proceedings of the 1st IEEE Conference on Control Application.* Dayton, OH. pp. 938–943.

Zhang, Y. M., B. L. Walcott and L. Wu (1992b). Dynamic modeling of full penetration process in GTAW. In: *Proceedings of American Control Conference.* Vol. 4. Chicago. pp. 3345–3349.

Zhang, Y. M., H. E. Beardsley and R. Kovacevic (1994a). Real-time image process in 3d measurement of weld pool surface. In: *ASME International Mechanical Engineering Congress.* Vol. 68. pp. 255–262.

Zhang, Y. M., L. Li and R. Kovacevic (1995a). Dynamic correlation between weld shape and weld penetration. In: *ASME Interna-*

tional Mechanical Engineer Congress. Vol. 69. ASME. San Francisco, CA. pp. 883–898.

Zhang, Y. M., L. Li and R. Kovacevic (1995b). Monitoring of weld pool appearance for penetration control. In: *Proceedings of the ASM 4th International Conference on Trends in Welding Research* (H. B. Smartt, J. A. Johnson and S. A. David, Eds.). Gatlinburg, TN. pp. 683–688.

Zhang, Y. M., L. Li and R. Kovacevic (1997). Dynamic estimation of full penetration using geometry of adjacent weld pool. *ASME journal of Manufacturing Science and Engineering.*

Zhang, Y. M., L. Wu, B. L. Walcott and D. H. Chen (1993a). Determining joint penetration in GTAW with vision sensing of weld-face geometry. *Welding Journal* **72**, 463s–469s.

Zhang, Y. M., Liguo E. and B. L. Walcott (2001). Robust control of pulsed gas metal arc welding. *ASME journal of Dynamic Systems, Measurement and Control.*

Zhang, Y. M., R. Kovacevic and L. Li (1996a). Adaptive control of full penetration gas tungsten arc welding. *IEEE Transaction on Control Systems Technology* **4**(4), 394–403.

Zhang, Y. M., R. Kovacevic and L. Li (1996b). Characterization and real-time measurement of geometrical appearance of the weld pool. *International Journal of Machine Tools and Manufacturing* **36**(7), 799–816.

Zhang, Y. M., R. Kovacevic and L. Wu (1992c). Controlling welding penetration in GTAW using vision sensing and adaptive control technique. *ASME Transactions* **XX**, 317–324.

Zhang, Y. M., R. Kovacevic and L. Wu (1992d). Sensitivity of front-face weld geometry in representing the full penetration. *Proceedings of Institution of Mechanical Engineers, Part B: Journal of Engineering Manufacture* **206**(3), 191–197.

Zhang, Y. M., R. Kovacevic and L. Wu (1992e). Three-dimensional vision sensing weld penetration. In: *Proceedings of the IASTED International Conference on Control and Robotics.* Vancouver, Canada. pp. 301–304.

Zhang, Y. M., R. Kovacevic and L. Wu (1993b). Closed-loop control of weld penetration using front-face vision sensing. *Proceedings of Institution of Mechanical Engineers Part I: Journal of Systems and Control Engineering* **207**, 27–34.

Zhang, Y. M., R. Kovacevic and L. Wu (1993c). On-line measure of full penetration weld geometry. In: *Proceedings of the 12th World Congress of the IFAC.* Vol. 8. Sydney, Australia. pp. 97–100.

Zhang, Y. M., R. Kovacevic and L. Wu (1994b). Robust adaptive control of full penetration using weld depression feedback. In: *Proceedings of the Fifth International Welding Computerization Conference.* Golden, Colorado. pp. 99–110.

Zhang, Y. M., R. Kovacevic and L. Wu (1996c). Dynamic analysis and identification of gas tungsten arc welding process for full penetration control. *Transactions of ASME, Journal of Engineering for Industry* **118**, 123–136.

Zhang, Y. M., Y. X. Yao, L. Li and R. Kovacevic (1996d). Based robust multivariable linear control of weld pool shape. In: *Proceedings of the ASME Dynamic Systems Control Division.* Vol. 58. ASME International Mechanic Engineering Congress. Atlanta, GA. pp. 259–264.

Zhang, Y. M., Z. N. Cao and R. Kovacevic (1996e). Numerical analysis of fully penetrated weld pools in gas tungsten arc welding. *Proceedings of Institution of Mechanical Engineers, Part C: Journal of Mechanical Engineering Science* **210**(2), 187–195.

Zhang, Y.M., E. Liguo and R. Kovacevic (1998b). Active metal transfer control by monitoring excited droplet oscillation. *Welding Journal (Miami, Fla)* **77**(9), 388s–395s.

Zhu, P., M. Rados and S. W. Simpson (1995). A theoretical study of a gas metal arc welding system. *Plasma Sources, Science and Technology* **4**(3), 495–500.

Index

Adaptive Control, 165, 244
Adaptive Control: GTAW, 171
Aerodynamic Drag, 47
Aerodynamic Force, 47
ANN: Artificial Neural Networks, 55
Arc Blow, 22
Arc Characteristics, 20
Arc Column Voltage, 37
Arc Length, 41
Arc Length Control, 150
Arc Length Sensors, 103
Arc Voltage, 11, 41
Arc Welding, 2
Artificial Neural Networks, 55
Automatic Control Techniques, 148
Automatic Voltage Control, 184

Back-Face Sensing, 106, 107

Calibration, 223
Case Study: Control of GMAW, 219
CC: Constant Current, 64
CCT: Continuous Cooling Transformation, 20
Charge-Coupled Device, 97
Classical Control, 160
Classical Control: PI, 160
Classical Control: PID, 160
Classification of Sensors, 95

Computer-Based Measurements, 99
Constant Current: CC, 64
Constant Voltage: CV, 66
Contact Sensors, 96
Contact Tip-to-Workpiece Distance, 59
Continuous Cooling Transformation: CCT, 20
Control of GMAW: A Case Study, 219
Control: Cooling Rate, 152
Control: Droplet Transfer Frequency, 155
Control: GMAW, 147
Control: Joint Profile (Fill Rate), 158
Control: Joint Profile Trajectory, 158
Control: Mass and Heat Transfer, 151
Control: of Weld Geometry, 153
Control: Weld Penetration, 156
Control: Weld Pool, 153
Control: Weld Temperature, 152
Cooling and/or Solidification Rates, 20
Cooling Rate: Control, 152
Cost of GMAW, 57
CV: Constant Voltage, 66

Dawn of GMAW, 57
Detachment Criteria, 48
Diffusion Welding: DFW, 3
Direct Model Reference Adaptive Control: DMRAC, 220
Distributed Parameter Model, 54
Disturbance Rejection Test, 242
DMRAC: Direct Model Reference Adaptive Control, 220
DMRAC: Experimental Results, 262
DMRAC: Formulation, 257
DMRAC: Implementation, 260
Droplet Dynamics, 47
Droplet Transfer Frequency, 112
Droplet Transfer Frequency: Control, 155

Electro-gas Welding: EGW, 3
Electro-slag Welding: ESW, 3
Electrode Extension (Stick-Out), 41
Electrode Orientation, 42
Electron Beam Welding: EBW, 3
Empirical Models, 53
Empirical Modeling, 222
Empirical Transfer Function Model, 234
Environmental Issues, 186
Estimation, 54
Experimental Data, 223
Expert System, 176

Feedback Control Techniques, 148

Feedback Linearization, 181
Fill Rate Control, 158
Flux-Cored Arc Welding: FCAW, 2
Free-Flight Transfer, 23
Friction Welding: FRW, 4
Front- or Top-Face Sensing, 107
Fuzzy Logic, 55, 174

Gas Metal Arc Welding: GMAW, 2
Gas Tungsten Arc Welding: GTAW, 2
Globular Transfer, 24
GMAW: Conclusions, 275
GMAW: Control, 147
GMAW: Gas Metal Arc Welding, 2
GMAW: Modeling, 9
GMAW: Principle of Operation, 11
GMAW: Sensing, 95
GMAW: Survey, 4
Gravitational Force, 30, 31, 47
GTAW: Modeling, 34, 35

HAZ: Heat Affected Zone, 17
Heat Affected Zone: HAZ, 17
Heat and Mass Transfer, 38, 251
Heat Transfer or Flow, 15

Idaho State University: ISU, 219
Identification, 54
INEEL, 219
INEEL/ISU Model, 44
INEEL: Idaho National Engineering and Environmental Laboratory, 44

INDEX

Infrared Sensing, 109
Intelligent Control, 174
Intelligent Modeling, 55
Intelligent Sensing, 116
Inverters, 67
IPPC: In-Process Penetration Control, 153
ISU, 219
ISU Adaptive Control Scheme, 170
ISU: Idaho State University, 44, 219
Iterative Learning Control, 181

Joint Profile Trajectory: Control, 158
Joint Profile: Control, 158

Knowledge-Based System, 176

Laser (Range) Finders, 103
Laser Beam Welding: LBW, 3
Light and Spectral Radiation Sensors, 104
Linearization, 52, 248

MAG: Metal Active Gas, 2
Magnetic Fields, 22
Manufacturing, 277
Mass and Heat Transfer Control, 151
Melting Rate, 22
Metal Transfer Characteristics, 23
Metal Transfer Experiments, 29
Model Equations, 49
Model Linearization, 52
Model Reference Adaptive Control: MRAC, 165
Model Simplification, 248

Model Simplification, 52
Modeling: GMAW, 9
Momentum Force, 47
MRAC: Model Reference Adaptive Control, 165
Multi-Loop Control, 239
Multivariable Control, 162

Neuro-Fuzzy Logic Control, 174

Optical Sensors, 115
Optimal Control, 164
Optimization, 164
Oxyfuel Gas Welding: OFW, 3

Phase-Locked Loop, 185
Physics of Arc, 15
Physics of Metal Transfer, 30
Physics of Welding, 15
PI (Proportional-Integral) Control, 160
PI Controller: SISO Current Control, 235
PID (Proportional-Integral-Derivative) Control, 160
Plasma, 21
Plasma Arc Welding: PAW, 3
Polarity, 40
Power Supplies, 62
Process Variables, 39, 40
Process Voltages, 36
Pulsed Current, 67
Pulsed Current Transfer, 26
Pulsed Transfer, 23

Quality Control, 180
Quality Control: Sensors, 116
Quantitative Feedback Theory, 184

Radiographic Sensing, 108
Relative Gain Array Analysis, 240
Relative Gain Array: RGA, 182
Reset Conditions, 50
Resistance Welding RW, 3
RGA: Relative Gain Array, 182

Safety Issues, 186
Self-Regulation of the Arc, 11
Sensing: GMAW, 95
Sensors Classification, 95
Sensors for Line Following, 101
Sensors for Seam Tracking, 101
Sensors: Contact Type, 96
Sensors: Non-Contact Type, 96
Sensors: Quality Control, 116
Sensors: Statistical Process Control, 116
Sensors: Weld Penetration Control, 106
Sensors: Weld Pool Geometry, 113
SFBT: Static Force Balance Theory, 18
Shielded Metal Arc Welding: SMAW, 2
Shielding Gases, 42
Short-Circuiting Transfer, 26
Single-Input, SinglE-Output: SISO, 235
SISO Current Control: PI Controller, 235
SISO: Single-Input, Singl-Output, 235
Smart Robotic Welders, 277
Solidification and/or Cooling Rates, 20
Spray Transfer, 23, 25

Stability Analysis, 256
Static Force Balance Theory: SFBT, 18
Statistical Models, 53
Statistical Process Control, 180
Statistical Process Control: Sensors, 116
Stick-Out (Electrode-Extension, 41
Stick-Out Voltage, 37, 38
Streaming Transfer, 26
Submerged Arc Welding SAW, 2
Surface Tension, 31, 32
Surface Tension Force, 47, 48
Survey: GMAW, 4

Temperature, 100
Thermit Welding: TW, 3
Trace Element Method, 110
Travel Speed, 41

Ultrasonic Sensing, 108
Ultrasonic Welding (USW), 4

Variable-Polarity Plasma Arc Welding, 184
Voltage Measurement, 103

Weld Bead Geometry, 32
Weld Geometry: Control, 153
Weld Penetration Control: Sensors, 106
Weld Penetration: Control, 156
Weld Pool, 32
Weld Pool Depression Method, 110
Weld Pool Geometry: Sensors, 113
Weld Pool Oscillation, 111

Weld Pool: Control, 153
Weld Temperature: Control, 152
Welding Current, 101
Welding Parameters Monitoring, 100
Welding: Space Research, 277
WFS: Wire Feed Speed, 40
Wire Feed Speed, 150